Swarm Intelligence and Machine Learning
Applications in Healthcare

Editors

Shikha Agarwal

Associate Professor, University Institute of Technology
Rajiv Gandhi Proudyogiki Vishwavidyalaya
Bhopal, India

Manish Gupta

Amity University
Madhya Pradesh, India

Jitendra Agrawal

Associate Professor, School of Information Technology
Rajiv Gandhi Proudyogiki Vishwavidyalaya
Bhopal, India

Dac-Nhuong Le

Associate Professor, Faculty of Information Technology
Haiphong University
Haiphong, Vietnam

Kamlesh Kumar Gupta

Rustamji Institute of Technology
BSF Academy Tekanpur
Gwalior, India

CRC Press
Taylor & Francis Group
Boca Raton London New York

CRC Press is an imprint of the
Taylor & Francis Group, an **informa** business

A SCIENCE PUBLISHERS BOOK

First edition published 2022
by CRC Press
6000 Broken Sound Parkway NW, Suite 300, Boca Raton, FL 33487-2742

and by CRC Press
4 Park Square, Milton Park, Abingdon, Oxon, OX14 4RN

© 2022 Taylor & Francis Group, LLC

CRC Press is an imprint of Taylor & Francis Group, LLC

Library of Congress Cataloging-in-Publication Data (applied for)

ISBN: 978-1-032-14579-2 (hbk)
ISBN: 978-1-032-14582-2 (pbk)
ISBN: 978-1-003-24003-7 (ebk)

DOI: 10.1201/9781003240037

Typeset in Times New Roman
by Radiant Productions

Preface

The Healthcare Sector is characteristic of complexities, dynamism and variety. In the 21st century, the healthcare domain faced huge challenges in terms of disease detection and prevention, high costs, skilled work force and better infrastructure. In order to handle these challenges, Intelligent Healthcare management technologies are required to play an effective role in improving a patient's life. Healthcare organizations also need to continuously discover useful and actionable knowledge to gain insight from tonnes of data for various purposes to save lives, reduce medical operation errors, enhance efficiency, reduce costs and making the whole world a healthy place.

This unique book introduces a variety of techniques designed to represent, enhance and empower multi-disciplinary and multi-institutional machine learning research along with swarm intelligence in healthcare informatics. This book also discusses one of the major applications of artificial intelligence: the use of machine learning to extract useful information from multimodal data in an optimized way by using swarm intelligence techniques. It discusses the optimization methods that help minimize the error in developing patterns and classifications, which further helps improve prediction and decision-making.

Applying Swarm Intelligence and Machine Learning Techniques in Healthcare is essential nowadays. The objective of this book is to highlight various Swarm Intelligence and machine learning techniques for several medical issues in terms of Cancer Diagnosis, Brain Tumor, Diabetic Retinopathy, Heart disease as well as drug design and development. This book will act as a one-stop reference for readers to think and explore Swarm Intelligence and machine learning Algorithms seriously for real-time patient diagnosis, as it provides solutions to various complex diseases found critical for medical practitioners to diagnose in the real-world.

Contents

CHAPTER 1
An Intelligent Methodology for COVID-19 RISK Prediction using Swarm Intelligence Optimization
A Machine Learning Perspective

V Kakulapati[1],* and *Sheri Mahender Reddy*[2]

1. Introduction

Two hundred twenty countries throughout the world have been affected by COVID-19. Countries like the US, India, Brazil, and Russia, with their large population, are seriously impacted. RT-PCR-based validity and reliability testing for COVID-19 identification for a large percentage of the population in these regions is a severe threat. A COVID-19 Epidemic was announced by the World Health Organization in Wuhan, China, in December 2019 [1], leading to almost 4,066,041 fatalities and 188,626,836 million diagnoses globally at the time of this writing. Initial testing, identification, and assessment are essential in assisting medical practitioners in protecting victims within this epidemic. Numerous investigations utilize Artificial Intelligence (AI) methodologies to enhance various processes regarding efficiency, precision, and performance in a health context.

[1] Sreenidhi Institute of Science and Technology, Yamnampet, Ghatkesar, Hyderabad, Telangana-501301.
[2] Otto-Friedrich University of Bamberg, IsoSySc.
Email: mahender-reddy.sheri@stud.uni-bamberg.de
* Corresponding author: vldms@yahoo.com

The disease is triggering COVID-19, namely SARS-CoV-2 variants. The majority of modifications have little or no effect on the characteristics of the virus. However, several changes may affect the features of genomics, such as ready transmission, associated significant illnesses or vaccine efficacy, screening methods and testing technology [2].

As the infection is widespread throughout a population, the chance of viral mutations increases, leading to numerous diseases. More possibilities that a pathogen will reproduce will multiply where there are more chances of mutations. In this investigative work, the mortality risk factor after 19 distinct variants of Covid-19 have been thoroughly analyzed, B.1.617, B.1.618, B.1.617.1, and B.1.1.7 are some of the common variations worldwide. The forecasting framework is designed for utilizing several classifications such as decision trees, Ada boost, Gaussian Naive Bayes, Random Forest, ANN gradient descent, and to evaluate covid 19 patients' mortality risk. The Particle Swarm Optimizing techniques were then applied to enhance forecasts. Because of the current crisis worldwide, this model will help the patients decide on treatment and use technology to manage emergencies.

For patients with pre-existing comorbidity diseases, such as Mellitus, cardiovascular problems, pulmonary ailments, digestive and kidney diseases, inflammatory infections, rheumatoid abnormalities and, obesity, their COVID-19 prognosis became worse. Also, it was concluded that COVID-19 can, in various instances, result in multiple multi-organ dysfunction leading to severe conditions.

The influences of pandemic transmission may be significantly reduced using AI (Artificial Intelligence) and ML (machine learning) technologies [3–5]. ML patient information integration methods come under a variety of possible investigation guidelines [6]. Various researchers have demonstrated the relevance of machine learning, particularly for forecasting. Most of the investigations proposed to anticipate and forecast, but additional analysis is still necessary, and discoveries about COVID-19 improve with a reliable collection of medical records.

AI technology is Intended to determine COVID-19 risk analysis. ANN, which may be employed alone or in combination with other statistical approaches, is among the most popular and frequent methods in use [7]. ANN cannot understand the absolute consistency with the appropriate target attribute in complicated nonlinear models such as those for infectious diseases.

A new COVID-19 risk prediction system is building, improving decision support classification precision, and reducing computer costs or functionalities. The method needed to enable decisions is GNB (Gaussian Naive Bayes). In its approach GNB employs the mathematical model based on the current assessment performance to evaluate features. The critical search strategy, is to get an optimal amount of ranking characteristics, which each of the feature subsets developed accomplishes with the GNB model, utilized as a ML classification. It is important to mention that the GNB technique includes a basic forecasting model, yet it is more effective than more sophisticated forecasting analytics, like ANN.

The IDSS (Intelligence Decision Support system) is often used in enterprises, farming, transport, environmental protection, and other areas [8]. It plays a pivotal role in a method that helps social change. There are several challenges with disease screening: disease risk predictions, diagnoses of infectious diseases, and outcome forecasts [9]. Once the challenges have been thoroughly studied, it has been determined that the appropriate PSO (swarm intelligence optimization) algorithms are used to improve the analytical thinking of ML algorithms. The parameters of ML are divided into two parts: the search for solutions and the development of evolutionary algorithms to ensure that challenges may be solved conveniently and intelligently [10].

One appropriate forecasting methodology is the ANN (artificial neural network). It is significantly more adaptable, and it can address challenging and inappropriate instances better than the methodology of recurrence. A wide variety of ANN methods rely on the forecasting techniques described, and one of them is BP (backpropagation). ANN classification was demonstrated in the COVID-19 Data set prognosis in [11], employing the BP technique for model development and integrating 13 clinical functionalities as intake and anticipating the absence or existence of coronavirus. Different methods used by ANN and several others have also been created with the help of Kaggle data and discovering pattern techniques, including DT (Decision Tree), NB (Naive Bayes), to evaluate and forecast their effectiveness and attain a F-measure. The intake ANN framework incorporates 13 risk factor characteristics, and the outcome is COVID-19.

The rest of the chapter is formulated as follows. Existing techniques are explained, and the reviewed approaches are discussed in Section 2. The risk related to COVID-19 is explained in Section 3. Section 4 addresses the methodology which is implemented in this work. The implementation setup is addressed in Section 5. In Section 6, the performance of the proposed method is discussed in detail. Section 7 exhibits the concluding remarks of the chapter, and Section 8 describes the future research directions followed by a reference section.

2. Related Work

Scientific systems have increasingly utilized precise and personalized impact assessment, generally measured by logistic forecasting models [12]. Because of their multivariate properties, these approaches incorporate various interactions of multiple variables and are suitable for the risk analysis process. This methodology includes confidential data, often obtainable through observational electronic medical records [13] or prospective datasets developed for scientific objectives [14].

Numerous reports see the worsening effects of severe abnormalities, including pulmonary disorders, in infected patients [15–19]. Because of COVID-19's intensifying and increasing mortality by many susceptible populations, research on epidemiology of the novel virus, such as the recognition of factors associated with significant fatalities and morbidity, was to become more necessary in all probability [20]. The history of significant disease conditions in individuals in with COVID19 was in [21]. A context demonstrated that the considerable risk characteristics for moderate COVID19 illness were high blood pressure, Mellitus, pulmonary disease, coronary heart disease, and neurological disease [22]. Lockdowns have been established as a cause for depression [23]. There is good evidence of psychiatric disorders related to the previous SARS epidemic [24]. Outbreaks generate nervousness and sadness.

An investigation [25] similarly utilized SVM to categorize the intensity of signs in the COVID-19 patient and used it for binary decision data comprising outcomes from urinalysis testing and combined patients with moderate symptoms with chronic conditions. The analysis indicates that variables with increased precision had a solid link to acute COVID-19. This data demonstrated a high-performance association with acute COVID-19. It should be recognised that the intensity of instances, ranging from severe to mild, is largely influenced by a range of factors. There were more severe instances in adults aged 65 years, unlike other patients. In addition, the risks of experiencing acute COVID-19 are much more significant in adult patients. The medical evidence in the collections of body fluids indicates statistically considerable variation from the outcomes of tests in moderate/severe instances.

A significant risk indicator of higher intensity is identified in the context of an underlying serious illness [26, 27]. According to the past findings, the prevalent elevated blood pressure, Cardiac issues, acute kidney problems, and cancer in COVID-19 sufferers were substantially related to higher treatment response and high mortality rate. No correlations identified chronic Liver Disease and COVID-19 inclination and possibly due to a massive conserving efficiency viral onslaught [28] and good hepatic resistance and regenerating ability [29].

One of the main components of controlling the medical system throughout the high-demand periods is the prediction of chronic illness [30]. It also assists in determining the early need and provides for the treatment plan in advance. A more robust cerebral network is becoming a forecast of the people affected. The ML method is designed to identify COVID19 genomes [31]. The authors presented a decision tree technique, and the publicly accessible COVID19 data with the alfa coronavirus and Beta corona Virus and measured performance. The precision of categorization, accuracy, among others, are considering for evaluations. The criticality and surviving possibilities of patients suffering from COVID 19 infections are based on various health risks.

3. About COVID 19 Risk Prediction

The influencing factors are determined to correlate with traditional healthcare pathological knowledge for acute COVID-19. It is common to understand whether aging is a crucial vulnerability indicator for chronic COVID-19 [32]. The involvement of comorbidity and the reduced immune system effectiveness in connection with normal aging include this [33, 34]. Obesity is a risk factor for developing complications like high blood pressure, heart problems, and cancer Overweight or obese. Nevertheless, metabolism has significant consequences, along with a boost in inflammatory circulation [35]. Patients and healthcare professionals with health disorders must collaborate and address these issues with regard and certainty [36]. The steps to be taken by people regarding health problems as well as other chronic diseases:

3.1 Prescribed Drugs Persist Daily

The patients suffering from chronic diseases such as pulmonary issues, liver problems, cardiac issues, irregular physical exercise and food intake, and diabetes on mediation and treatment do not stop their previous prescribed medicines. Throughout this epidemic, the infected patients may experience a strong impact. Panic and nervousness may be prevalent and generate emotional responses. In elderly persons with several other underlying medical conditions—like cardiovascular disease, impaired respiratory system, obesity, or Mellitus, the probability of developing catastrophic indications of COVID-19 can be significant. It's analogous to how most respiratory disorders, such as pneumonia, are experienced. Although each such situation can raise the likelihood of acute COVID-19 indications, those with numerous other health issues are considerably more in danger.

Identifying the appropriate indications for negative impacts might alter the essential findings and the virus phase. Therefore, evaluating a ML predictive model of metabolic abnormalities, SARS-CoV-2 might leverage the established methods to determine morbidity and mortality variables, including comorbidity, diagnostics, and essential indicators while patients progress along the pathway, i.e., Diagnosis, hospitalization, and, when appropriate, briefly pre and post-entry to the Intensive Care Unit [37]. A COVID-19 member also increases the possibility of significant peer group assessments. This probability is accurate with the aggregation of parental diseases and exposes inadequacies of the infection containment policy focusing upon this domestic lock-down or distancing the symptomatic.

3.2 Patient Way of Life

There is evidence of changed accommodation, shifts in average earnings and social standing due to the COVID-19 pandemic among, affected people in the population forcing the enforcement of public sector interventions in the form of semi and full lock-downs to exercise biosafety strategies during the early disease outbreak [38].

Guidelines of the epidemic, including social distancing, resulting in lasting loneliness, increased panic, depression, and melancholy, are influenced by the COVID-19 behavioral health conditions [39, 40] that could alter daily routine behavior. The prevalence of risks during this stage in addition cause psychological distress due to epidemiological fear.

3.3 Parameters of Patient Sustainability

For providing security to the elevated coronaviral populations a reliable approach is to restrict travel during the pandemic. This process involves developing 'green zones' enabling elevated patients to stay protected and comfortable, either within their family or communities, for a considerable time during such an epidemic which investigates the effectiveness of precautions to protect people at a significant risk of severe consequences due to COVID-19. This protection strategy supports good epidemiologic concepts to reduce the frequency and duration of effective interactions and the risk of infection among high-risk groups [41]. However, efficient protection techniques must be adequately developed and adapted to fit the diverse contexts [42].

3.4 Higher Incidence of Consequences

- Severe lung disorders, including sinusitis, allergies
- Cardiovascular diseases
- Kidney related diseases
- Hepatitis
- Acute neurological diseases
- High blood glucose levels and blood pressure
- Diverse cancer chemotherapy such as breast, skin, and lung
- Obesity
- Pregnancy [43]
- Body part transplantation
- COPD (chronic obstructive pulmonary disorder)

- Deficient of immune system
- Colorectal surgery with bowel resection
- Knee replacement

3.5 Usage of Drugs

Numerous drugs were employed worldwide for COVID-19: chloroquine and hydroxychloroquine, favipiravir, tocilizumab, lopinavir/ritonavir, and azithromycin, convalescent plasma treatments, vitamin C, antiviral and gout steroids, and remdesivir were authorized by the FDC for treating COVID-19 [44, 45]. COVID-19 may impact several body parts, such as pulmonary, cardiac, induced hepatic, kidney, immunological, and metabolic processes. Security issues, which psychologists are familiar with, i.e., the idea of when stimulant medications are prescribed on COVID-19 diagnosis pharmacological abnormalities can result in a significant disruption in these systems [46]. Although this condition has emerged and most participants indicated it is not recognized, adverse medication responses may be missing, and particular care is requested. Observing patient behavior should be considered to increase the lifespan of chronic psychological patients with COVID-19.

3.6 Risks are Associated with Travelling

No travel is allowed for identified, probable, and suspected Covid-19 patients and their connections. Patients are in isolation. Containment of interactions from these cases is a must. Individuals with any COVID-19-consistent indication or symptoms do not fly without a Covid-19 diagnostic to prevent the spread of SARS-CoV-2 disease. A postponement of travel is recommended for those at a risk of catching severe illnesses and death, including people aged 60 with underlying conditions since they are at a greater risk of Covid-19 severity.

4. Methodology

The intelligent framework comprises the decision-making model which uses binary classification methods to identify covid-19 risk assessment. The framework is used to identify the indicators and attempts to detect the illness. Rather than a lead to a different procedure, a conventional method uses if and then-else logic. In this chapter, ML ensemble techniques build an intelligent system rather than rules that exist along with particle swarm intelligence optimization. Through many crucial steps, the intelligent system generates outcomes. Ensembles refer to the algorithm formed by combining several machine learning models into a single proficient model [18]. Ensemble algorithms are using to improve the accuracy by combining weak predicting models. The algorithm for the ensemble refers to the algorithm that combines many models into a single effective model [47]. Collectively these are used to enhance precision by integrating inadequate models to build a robust forecasting model.

4.1 AB (Ada Boost)

It represents an adaptive boosting algorithm and is the first practical supervised learning application [48]. It may be used as a set to several problems such as analysis, classifying, and employing an evolutionary strategy to repeatedly improve ineffective classification inaccuracies and probably make a reliable merged classification.

4.2 GD (Gradient Descent)

It is a methodology to significantly improve various optimization processes and mobility. The algorithm is employed based on a COVID-19 finding function. The descendant variables are usually selected by group, and that is the data set, rather than the whole dataset, for every repetition of the Stochastic gradient. The gradient computation of every repetition can be supporting by chosen batches [49].

4.3 RF (Random Forest)

Integrates the forest classifiers to allow the individual trees to be separated based on the measured probability variable of the array [50]. It utilizes a testing and training, classification model. It splits the data set into a subgroup, using random forests for decision-making. The forests incorporate a feature vector that is useful for predicting. The random forest optionally receives an approximation, i.e., the number of forests and the maximum characteristics, to be picked as hyper-parameters.

4.4 DT (Decision Tree)

The standard ML methodology for risk scores is a simple ensemble learning method [51, 52]. This methodology is accurate in classifying specific novel methods introduced [53, 54] and displays transparent interpretive findings. This technique nevertheless provides a performance ratio if the sample is a time series with detailed categorization. It constructs a tree-shaped architecture as part of the descriptive method. Initially it is able to identify the optimal values of a particular indication. Subsequently, it splits every sub-sample with an excellent value till default stopping criteria have been achieving.

4.5 NB (Naive Bayes)

It is noteworthy for the accuracy and ease of various trained predictions. This approach understands the probability of an item with specific characteristics of a particular category or class. It is a collection of predictive statistical methods. Such methods are known as "naive" as they employ the naïve hypothesis to be independent. That is, the approaches assume that the existence of a characteristic is distinct from other specific factors. This model relies on the Bayes principle or criterion, which assesses whether a particular occurrence belongs to a particular category.

4.6 ANN

A network of several operational layers of associated translational linguistics or terminals processing elements. Every receptor mixes its signals proportionally and instead goes across a standard or multivariate filtering learning algorithm. Predictable processes combining takes place by summarizing weighting and incoming elements. It creates the objective via a feed-forward communication process and instead re-propagates mistakes throughout the learning step to upgrade every receptor's weight [55].

4.7 PSO (Particle Swarm Optimization)

This is one of the basic heuristics employed and an evolutionary computing process for a population size. A set of parameters has its quantities modified in several epochs nearer to the companion closest to the destination at any moment. This algorithm is intended to obtain the best optimal solution and is effectively used as the operation. Data is obtained solely through the assessment of a function. It's straightforward to execute. Probabilistic particles and exploration for an optimal solution always seem to be established for the PSO. Six significant parameters are analyzed:

- event measure
- gBest (Global best value) which particles are nearest to the target
- Data that is a viable solution
- whether the data modified
- pBest (personal best) value reflecting the nearest values to the target

5. Implementation Results

The data set is collecting from Kaggle for implementation. The data set comprises patient details of those infected with covid 19 (variant type) with varied chronic diseases history and post-recovery effects such as long-term symptoms of covid 19, fever, weakness, dizziness and fungal infection. In this work, supervised machine learning technology is employed where the model is trained on a labeled data set. This data is executed on different models like Decision tree, Adaboost, Random forest classifier, Gradient Descent, ANN, and Gaussian naïve Bayes algorithm with the same dataset. All the models are trained and tested. The accuracy of each model is used, and the model with the highest accuracy rate is selected for risk prediction and the consequences.

The risk assessment of individuals with distinct covid-19 variations of associated serious illnesses is conducted. The data set of the forecast was employed to achieve precise findings. In this section, the pre-processed dataset has been utilized to train many ML prediction methodologies for execution. The ADA Boost, Decision Tree, Gaussian naïve bayes, Random Forest, ANN, and Particle Swarm Optimization approaches were analyzed. AdaBoost is another inbuilt ensemble applied for making predictions for the risks associated with covid-19 variants and severe diseases.

Another appropriate prediction approach is the Artificial Neural Network (ANN). It is far more versatile and can manage more complicated and unexpected circumstances than the technique of correlation. It is a way to

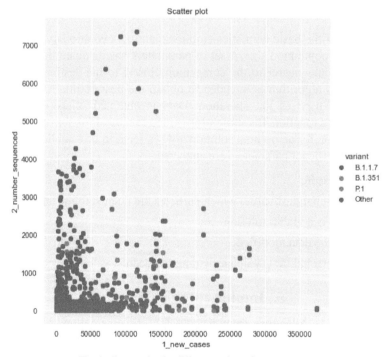

Fig. 1. Scatter plot for different variants for new cases.

address prediction issues with positive outcomes, although its effectiveness impacts the type of optimization employed in the workout. Generally speaking, the gradient descent technique is the optimization method. Neural network weights adjust by reducing the error between the neural network output and the actual value or target. The 'particle' swarm optimization technique optimizes the mistake. For the networking testing stage, the final weight is utilizing to assess the number of people suffering from COVID-19. The training data and testing data are using to validate the accuracy technique during the testing stage.

AdaBoost Classifier Model Accuracy: 74%

Model Accuracy with SVC Base Estimator: 83%

Array [135, 68, 51, 101],

[156, 46, 145, 15],

[134, 139, 18, 55]

Decision tree confusion matrix:

[[33   4]

[12   62]]

Accuracy: 85.5%

A sequence of hundred decision-tree classifications was employed as fundamental participants in the performance assessment of the Random Forest technique. The optimum division of every tree has been determined. AdaBoost was used for the group of classification trees. Each tree creates an ensemble according to a random parameter in the suggested arrangement. The final forecast is calculated by adding the output of each tree.

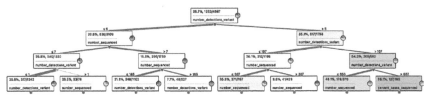

Fig. 2. Decision tree for 1139 nodes 570 leaves for covid-19 variant B.1351.

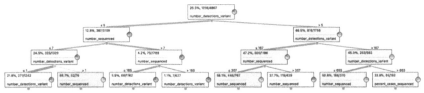

Fig. 3. Decision tree for 1139 nodes 570 leaves for covid-19 variant B.1.1.7.

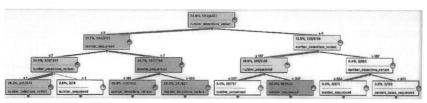

Fig. 4. Decision tree for 1139 nodes 570 leaves for covid-19 variant P.1.

Fig. 5. Decision tree for 1139 nodes 570 leaves for other.

Random forest confusion Matrix: [[29   8]
                                  [4   70]]

Accuracy: 89%

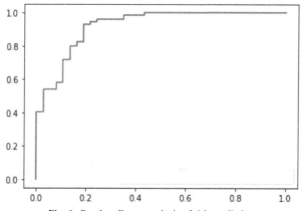

Fig. 6. Random Forest analysis of risk prediction.

Gaussian Naïve Bayes
Confusion matrix: [[32 5]
                   [10 64]]
Classification Accuracy: 86.4%

Results are calculated based on the number sequence and the variant sequence count. We get an optimum particle position and best error set as a risk limit of the variant, and other cases above the limit are treating as risky

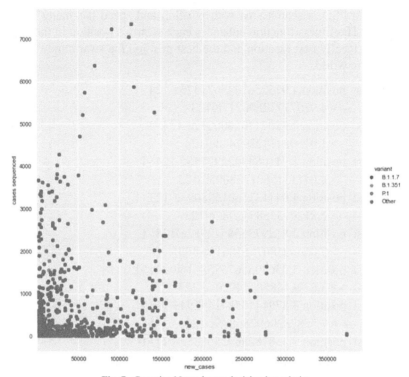

Fig. 7. Gaussian Naïve bayes decision boundaries.

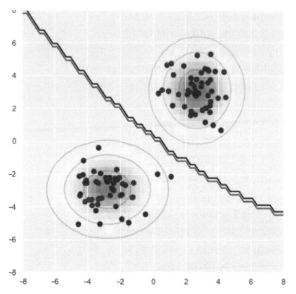

Fig. 8. Naïve bayes model for covid-19 variants risk.

variants. In PSO, a search area with position and speed has many swarms allocated. The fitness function enhances each swarm's position in the search area to its locally best location and the best position the swarm may achieve in the entire world.

global_best_position 4.9732561445621376e-124 -1
ERROR-------> 4.92077732800217e-123
global_best_position 4.92077732800217e-123 -1
ERROR-------> 1.915084342297448e-124
global_best_position 1.915084342297448e-124 -1
ERROR-------> 4.0411315601598095e-123
global_best_position 4.0411315601598095e-123 -1
ERROR-------> 2.812973558542982e-123
global_best_position 2.812973558542982e-123 -1
ERROR-------> 5.0568958675558616e-124
global_best_position 5.0568958675558616e-124 -1
ERROR-------> 3.8791448584149294e-123
global_best_position 3.8791448584149294e-123 -1
ERROR-------> 7.558782969256885e-124
global_best_position 7.558782969256885e-124 -1
ERROR-------> 8.575105763145442e-124
global_best_position 8.575105763145442e-124 -1
ERROR-------> 2.967412649518054e-123
global_best_position 2.967412649518054e-123 -1
ERROR-------> 2.2493461068681053e-123
And for seq and variants global
global_best_position 3.9368862688884314 -1
ERROR-------> 3.0629288500776637
global_best_position 3.0629288500776637 -1
ERROR-------> 3.5272180144054044
global_best_position 3.5272180144054044 -1
ERROR-------> 3.2607541782626224
global_best_position 3.2607541782626224 -1
ERROR-------> 2.7504069632585506
global_best_position 2.7504069632585506 -1
ERROR-------> 2.5513831548809156
global_best_position 2.5513831548809156 -1
ERROR-------> 2.7328791219834585
global_best_position 2.7328791219834585 -1
ERROR-------> 3.4898992309090504

global_best_position 3.4898992309090504 -1
ERROR-------> 3.4271937629475175
global_best_position 3.4271937629475175 -1
ERROR-------> 4.622581330374499
global_best_position 4.622581330374499 -1
ERROR-------> 2.9354159597670506
global_best_position 2.9354159597670506 -1
ERROR-------> 2.16746784274822

However, the global best position tends to be near zero as most variant number sequences compromise with new cases. Adding additional data with new features may overcome this situation.

Ada boosting
Out [12]:
new_cases number sequenced ... number_detections_variant variant

	new_cases	number sequenced	...	number_detections_variant	variant
0	5152	4.0	...	0	B.1.1.7
1	5152	4.0	...	0	B.1.351
2	5152	4.0	...	0	P.1
3	5152	4.0	...	4	Other
4	7365	24.0	...	0	B.1.1.7
...	...	...	...	...	...
5178	41170	1167.0	...	30	Other
5179	41065	731.0	...	700	B.1.1.7
5180	41065	731.0	...	15	B.1.351
5181	41065	731.0	...	0	P.1
5182	41065	731.0	...	16	Other

[5183 rows × 5 columns]
0   B.1.1.7
1   B.1.351
2   P.1
3   Other
4   B.1.1.7
Name: variant, dtype: object

A classification technique such as AdaBoost involves the pre-processed data on which the decision tree, random forest, gradients, and Gaussian Naïve Bayes are used to detect prediction accuracy. By applying several classification methods and particle swarm optimization, the learned correctness of this system is considerably enhanced. The model helps physicians forecast and diagnose covid 19 severity correctly using a selection of capabilities.

6. Discussion

Predicting severity and fatality assists in prioritizing those at significant risk, providing them with the best necessary care, and possibly increasing better results. It can also lower pressures on the health sector, encourage decision-making, and make good use of resource constraints. The observations lead to the conclusion that the effect of a new SARS-CoV-2, which is more transmissible than the earlier versions, will lead to more significant difficulties due to the COVID-19 pandemic under some probable conditions. There are many options for the variants of SARS-CoV-2 results.

Complications are sometimes correlated to or recognized with the underlying disease; people who are already unwell for another reason may have greater chances of requiring hospitalization or surviving the virus. Improving the quality of medication might also affect results. Aging is a risk of chronic diseases that can have a substantial effect on outcomes. Genomic variable hosting is probably also important, and new data shows that the COVID-19 observations impact two neighboring sensitivities [56]. These findings imply that medical practitioners should raise the frequency or severity of disease to prevent more infected persons, admissions, and fatalities. The results indicate that there would be more fatalities, even if the new SARS-CoV-2 variants are not any more severe than the prior.

PSO has a greater rate of convergence but seems to be having a problem with overfitting. The suggested swarm intelligence forecast algorithm may anticipate the hazards related to various versions of Covid 19 and the risk analysis of individual patients, which helps to improve the risk's forecast precision. To explain the possibility of picking additional hyperparameters like iteration and batch size, we expand the suggested ANN-PSO. If we use the same activation function for all the levels and use various activation functions for every layer, we analyze the influence of the activation function.

The performance level of this work is pretty much due to accuracy performance. The suggested technique should not be utilized for any strategy or intention changes but for the enhancement of awareness and forecasts. The predictions' accuracies also supports other aspects, including modifications—incorrect reporting of information, changes to the data analysis guides and methodological limitations. The approaches of optimizing particulate swarms are highly likely to anticipate the COVID-19 risk analysis.

7. Conclusion

In Outbreak patient populations with severe ailments and the association of increased incidence with the dysfunction of vital organs, the thorough monitoring and management of these elderly communities, in specific COVID-19 cases with severe illnesses and their acceptance, must be based upon the elevated risk of severe morbidities mortality. The risk of developing extreme risk in individuals, old, and fatty comorbidity, exceptionally high blood pressure, Mellitus, or Cardiovascular, was more significant.

8. Future Scope

The sequence of events with COVID-19 is highly traumatic, since it is challenging to foresee the way events evolve and circumstances change fast. However, certain aspects cannot control this situation, which is the case across several aspects of people's lives. This incorporates others' behaviors and emotions, whether the condition is prolonged and lasts long or If it can recur in the future. The enhanced training and learning method significantly improve the prediction accuracy of Covid-19.

References

[1] Zhu Na et al. 2020. A novel coronavirus from patients with pneumonia in China, 2019. N. Engl. J. Med. 382(8): 727–33.

[2] https://www.who.int/en/activities/tracking-SARS-CoV-2-variants.

[3] Albahri, A. S. R. et al. 2020. Role of biological data mining and machine learning techniques in detecting and diagnosing the novel Coronavirus (COVID-19): a systematic review. Journal of Medical Systems 44(7): 122.

[4] Zagrouba, R. et al. 2021. Modelling and simulation of COVID-19 outbreak prediction using supervised machine learning. Computers, Materials & Continua 66(3): 2397–2407.

[5] Kumar, A. et al. 2019. Anxious depression prediction in real-time social data. pp. 1–7. *In*: Proceedings of the International Conference on Advances in Engineering Science Management & Technology (ICAESMT), Dehradun, India, July 2019.

[6] Lalmuanawma, S. et al. 2020. Applications of machine learning and artificial intelligence for Covid-19 (SARS-CoV-2) pandemic: a review. Chaos, Solitons & Fractals 139: 110059.

[7] Thakkar, A. et al. 2021. Fusion in stock market prediction: a decade survey on the necessity, recent developments, and potential future directions. Information Fusion 65: 95–107.

[8] Nemati, H. R. et al. 2002. Knowledge warehouse: an architectural integration of knowledge management, decision support, artificial intelligence and data warehousing. Decision Support Systems 33: 143–161.

[9] McIntosh, B. S. J. et al. 2011. Environmental decision support systems (EDSS) development–challenges and best practices. Environmental Modelling & Software 26: 1389–1402.

[10] Al-Saedi, A. K. Z. et al. 2016. Improved multi join query optimization for RDBMS based on swarm intelligence approaches. Advanced Science Letters 22: 2847–2851.

[11] Karaylan, T. et al. 2017. Prediction of heart disease using neural network. Int. Conf. Comput. Sci. Eng. (UBMK) Antalya 2017: 719–723.

[12] Chen, J. H. et al. 2017. Machine learning and prediction in medicine—beyond the peak of inflated expectations. N. Engl. J. Med. 376: 2507–2509.

[13] Hippisley-Cox, J. et al. 2009. Predicting risk of osteoporotic fracture in men and women in England and Wales: prospective derivation and validation of QFractureScores. BMJ 339: b4229.

[14] Kanis, J. A. et al. 2008. The assessment of fracture probability in men and women from the UK. Osteoporos. Int. 19: 385–397.

[15] Huang, C. et al. 2020 Feb. Clinical features of patients infected with 2019 novel coronavirus in Wuhan, China. The Lancet 395(10223): 497–506.

[16] Guan, W. et al. 2020 Apr 30. China medical treatment expert group for Covid-19. Clinical characteristics of coronavirus disease 2019 in China. N. Engl. J. Med. 382(18): 1708–1720.

[17] Chen, N. et al. 2020 Feb. Epidemiological and clinical characteristics of 99 cases of 2019 novel coronavirus pneumonia in Wuhan, China: a descriptive study. The Lancet 395(10223): 507–513.

[18] Wang, D. et al. 2020 Mar 17. Clinical characteristics of 138 hospitalized patients with 2019 novel coronavirus–infected pneumonia in Wuhan, China. JAMA 323(11): 1061.

[19] Shi, H. et al. 2020 Apr. Radiological findings from 81 patients with COVID-19 pneumonia in Wuhan, China: a descriptive study. The Lancet Infectious Diseases 20(4): 425–434.

[20] Xu, X. et al. 2020 Feb 27. Clinical findings in a group of patients infected with the 2019 novel coronavirus (SARS-Cov-2) outside of Wuhan, China: retrospective case series. BMJ 368: m792.

[21] Lipsitch, M. et al. 2020 Mar 26. Defining the epidemiology of Covid-19—studies needed. N. Engl. J. Med. 382(13): 1194–1196.

[22] Zhao, D. et al. 2020. A comparative study on the clinical features of COVID-19 pneumonia to other pneumonia. Clin. Infect. Dis. 71(15): 756.

[23] Wang, B., Li, R., Lu, Z. and Huang, Y. 2020. Does comorbidity increase the risk of patients with COVID-19: evidence from a meta-analysis. Aging (Albany NY) 12(7): 6049–57.

[24] Carvalho, P.M.M. et al. 2020. The psychiatric impact of the novel coronavirus outbreak. Psychiatry Res. 286: 112902.

[25] Maunder, R. et al. 2003. The immediate psychological and occupational impact of the 2003 SARS outbreak in a teaching hospital. CMAJ 168(10): 1245–51.

[26] Yao, H. et al. 2020. Severity detection for the coronavirus disease 2019 (COVID-19) patients using a machine learning model based on the blood and urine tests. Frontiers in Cell and Developmental Biology 8: 1–10.

[27] Chan, J. W. et al. 2003. Short term outcome and risk factors for adverse clinical outcomes in adults with the severe acute respiratory syndrome (SARS). Thorax 58(8): 686–689.

[28] Booth, C. M. et al. 2003. Clinical features and short-term outcomes of 144 patients with SARS in the greater Toronto area. JAMA 289(21): 2801–2809.

[29] Yang, J. K. et al. 2010. Binding of SARS coronavirus to its receptor damages islets and causes acute diabetes. Acta Diabetologica 47(3): 193–199.

[30] Van Haele, M. et al. 2019. Human liver regeneration: An etiology dependent process. International Journal of Molecular Sciences 20(9): 2332.

[31] Krishna, L. et al. JAN, 2018. Early detection of peak demand days of chronic respiratory diseases emergency department visits using artificial neural networks. IEEE J. Biomed. Health Inform. 22(1): 285–290.

[32] Yan, L. H.-T. et al. 2020. Prediction of criticality in patients with severe Covid-19 infection using three clinical features: a machine learning-based prognostic model with clinical data in Wuhan. medRxiv.

[33] Randhawa, G. S. et al. 2020. Machine learning using intrinsic genomic signatures for rapid classification of novel pathogens: a COVID-19 case study. PLoS One 15(4): e0232391.

[34] Mahase, E. 2020. Covid-19: What do we know about "long covid"? BMJ 370. pmid: 32665317.

[35] Mahase, E. 2020. Covid-19: Why are age and obesity risk factors for serious disease? BMJ 371. pmid: 33106243.

[36] https://www.CDC.gov/coronavirus/2019-ncov/need-extra-precautions/people-with-medical-conditions.html.

[37] Zachary, Z. et al. 2020. Self-quarantine and weight gain related risk factors during the COVID-19 pandemic. Obes. Res. Clin. Pract. 14(3): 210–216. pmid: 32460966.

[38] Wang, C. et al. 2020. A longitudinal study on the mental health of the general population during the COVID-19 epidemic in China. Brain Behav. Immun. 87: 40–48. pmid: 32298802.

[39] Ahmed, N. J. et al. 2020. The anxiety and stress of the public during the spread of novel coronavirus (COVID-19). Journal of Pharmaceutical Research International 32(7): 54–59.

[40] Mossong, J. et al. 2008. Social contacts and mixing patterns relevant to the spread of infectious diseases. PLoS Med. 5(3): e74.

[41] Butler, N. et al. 2020. Considerations and principles for shielding people at high risk of severe outcomes from COVID-19: Social Science in Humanitarian Action Platform.

[42] https://digital.nhs.uk/coronavirus/shielded-patient-list/risk-criteria.

[43] Bilbul, M. et al. 2020. Psychopharmacology of COVID-19. Psychosomatics 61(5): 411–427.

[44] Bishara, D. et al. 2020. Emerging and experimental treatments for COVID-19 and drug interactions with psychotropic agents. Ther. Adv. Psychopharmacol. 10: 2045125320935306.

[45] Wang, T. et al. 2020. Comorbidities and multi-organ injuries in the treatment of COVID-19. Lancet 395(10228): e52.

[46] Jimenez-Solem, E. C. et al. 2021. Developing and validating COVID-19 adverse outcome risk prediction models from a bi-national European cohort of 5594 patients. Sci. Rep. 11: 3246.

[47] Dietterich, T. G. 2002. Ensemble learning. The Handbook of Brain Theory and Neural Networks 2: 110–125.

[48] Liu, H. et al. 2015. Comparison of four Ada boost algorithm-based artificial neural networks in wind speed predictions. Energy Conversion and Management 92: 67–81.

[49] Almustafa, K. M. 2020. Prediction of heart disease and classifiers' sensitivity analysis. BMC Bioinf. 21(1): 1–18.

[50] Liaw, A. et al. 2002. Classification and regression by random forest. R News 2(3): 18–22.

[51] Wang, G. et al. 2011. A comparative assessment of ensemble learning for credit scoring. Expert Systems with Applications 38(1): 223–230.

[52] Hoffmann, F. et al. 2007. Inferring descriptive and approximate fuzzy rules for credit scoring using evolutionary algorithms. European Journal of Operational Research 177(1): 540–555.

[53] Louzada, F. et al. 2011. Poly-bagging predictors for classification modelling for credit scoring. Expert Systems with Applications 38(10): 12717–12720.

[54] Zhang, D. et al. 2010. Vertical bagging decision trees model for credit scoring. Expert Systems with Applications 37(12): 7838–7843.

[55] Alić, B. L. et al. 2017. Machine learning techniques for classification of diabetes and cardiovascular diseases. 2017 6th Mediterranean Conference on Embedded Computing (MECO), pp. 1–4.

[56] Ellinghaus, D. et al. 2020. Genomewide association study of severe Covid-19 with respiratory failure. N. Engl. J. Med. 383(16): 1522–1534.

CHAPTER 2
Impact of COVID Vaccination on the Globe using Data Analytics

Pawan whig,[1,]* *Arun Velu,*[2] *Rahul Reddy*[3] and
Pavika Sharma[4]

1. Introduction

As the flare-up of the COVID-19 virus has become an overall pandemic, the ongoing investigations of epidemiological information are expected to set up the general public with better activity plans against the infection. Since the introduction of novel COVID-19 [1], the world is fretfully battling with its motivation. As of April 15, 2020, in light of the internationally shared live information by the Johns Hopkins dashboard, overall there are 138,976,244 affirmed cases, out of which 114,490,676 are recovered and 2,988,801 lost their lives [2]. Coronavirus has a place with the group of the SARS-CoV and MERS-CoV, where it starts with the underlying level indications of the common cold to serious degrees of respiratory illnesses causing trouble in breathing, sleepiness, fever, and dry hack [3]. Prasad et al. [4] saw that the identification of the infection can be improved by imaging utilizing immunoelectron microscopy strategies [5]. Till date, itemized morphology and ultrastructure of this infection remain not completely comprehended, and there are no explicit antibodies or medicines for COVID-19. Nonetheless, numerous continuous clinical preliminaries are assessing likely medicines.

[1] Senior IEEE Member, Dean Research Vivekananda Institute of Professional Studies, New Delhi, India.
[2] Researcher, Director Equifax, Atlanta, USA.
[3] Senior IEEE Member, University of Cumbersome, USA.
[4] Professor, BPIT, Rohini Delhi.
* Corresponding author: pawanwhig@gmail.com

Man-made reasoning (AI) can help us in dealing with the issues that should be tended to raised by the COVID-196 pandemic. It isn't just the advancement, in any case, that will influence the process but rather the data and imaginativeness of the individuals using it will play a role. In actuality, the COVID-19 crisis will most likely reveal a part of the critical deficiencies of AI. AI (ML), the current sort of AI, works by perceiving plans in chronicled prepared data. Individuals have a favored situation over AI. We can take in practices from one circumstance and apply them to novel conditions, drawing on our dynamic data to make the best hypotheses on what may work or what may happen. PC based knowledge systems, then again, need to pick up without any planning at what-ever point the setting or task changes even insignificantly.

A wide family of distinct viruses are coronaviruses. Any of them trigger the common cold in humans. Others, infected include bats, camels, horses, and wildlife. But how did SARS-CoV-2, the new COVID-19-causing coronavirus, come into being? SARS-CoV-2 emerged in bats, researchers say. That's also how the Middle East respiratory syndrome (MERS) and extreme acute respiratory syndrome (SARS) coronaviruses began [6]. At one of Wuhan's open-air "wet markets," SARS-CoV-2 made the transition to humans. It's where consumers buy fresh meat and fish, including animals that are slaughtered on the spot. Any wet market, includes cobras, wild boars, and raccoon dogs, sell wild or banned animals [7]. Crowded environments can cause genes to be swapped by viruses from various species [8]. The virus often changes so much that it may begin to affect and propagate among individuals [9]. It affected people who had no close contact with wildlife, as SARS-CoV-2 spread both within and outside China. This meant that the virus was spread from one person to the next. In the U.S. and across the world, it is already spreading, suggesting people are inadvertently receiving and sharing the coronavirus. This rising dissemination worldwide is what is becoming a pandemic [10].

In 1965, scientists first observed a human coronavirus [11]. This produced a widespread cold. Researchers identified a group of related human and animal viruses later that decade and named them for their crown-like shape. The Origin of how it can be transmitted is shown in Fig. 1

They will infect humans with seven coronaviruses [12–15]. The one that causes SARS established in 2002 in southern China and spread rapidly to 28 other countries. By July 2003, more than 8,000 people were sick, and 774 died. In 2004, a minor epidemic only affected four additional cases [15–19]. This coronavirus causes problems with fever, headache, and coughing, including cough and shortness of breath.

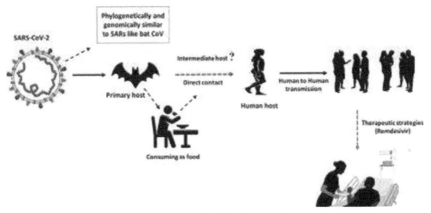

Fig. 1. Proposed origin of corona virus.

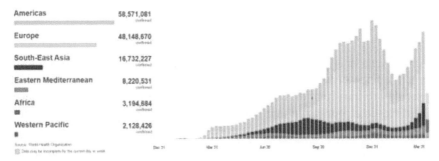

Fig. 2. Stats of COVID till April 2021.

As per April 2021 the total number of COVID cases were 138,976,244 with confirmed deaths of 2,988,801 with total 223 countries being infected as reported by WHO [20–25]. The stats of COVID are shown in the Fig. 2.

2. Data Analysis using Machine Learning

Machine learning is a data processing tool that automates the creation of analytical models. It is a subset of artificial intelligence focused on the premise that, with minimal human interaction, computers can learn from data, recognize trends and make decisions. Machine learning today is not like machine learning in the past, thanks to emerging computing developments [14]. It was born from the identification of patterns and the idea that computers would learn to do complex tasks without being programmed; academics interested in artificial intelligence decided to see how computers could learn from results [15]. The iterative nature of machine learning is important because it is able to adapt independently when models are introduced to new data [16–20]. To generate accurate, repeatable decisions and outcomes, they learn from previous

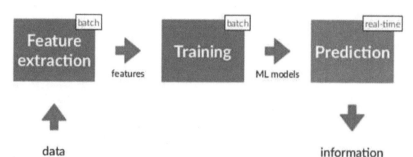

Fig. 3. Block diagram of process involved.

computations. It is a science that is not new, but has obtained new traction. In these research studies we used various machine learning libraries like Pandas, NumPy, etc. for obtaining the impact of the COVID vaccination on the globe [28–29]. The block diagram of the process involved is shown in Fig. 3.

3. Result and Process Step Wise

For the data analysis Google Colab is used, "Colab" is a Google Analysis product, for short. Colab enables anybody through the browser to write and execute arbitrary python code, and is specifically well suited to computer learning, data processing and education. The detailed analysis is presented stepwise to understand the process in easy ways.

Step 1: Importing Packages

```
import pandas as
pd importnumpy as
np importplotly.express as
px importplotly.graph_objects as go
fromplotly.subplots import make_subplots
```

Step 2: Importing Dataset

```
data=pd.read_csv('country_vaccinations.csv')
data2=pd.read_csv('countries-aggregated.csv')
```

Step 3: Cleaning Dataset

```
data.dropna(subset=['daily_vaccinations'],inplace=True)
s=data['date'].str.split('-',expand=True)
data['Year']=s[0]
data['Month']=s[1]
data['Date']=s[2]
```

```
fig1=px.scatter_geo(data,        color='vaccines',        locationmode="ISO-3",
locations="iso_code",  opacity=0.6,  hover_name="iso_code",  size="daily_
vaccinations", projection='conic equal area', animation_group
="iso_code",color_continuous_scale='blackbody',

animation_frame="Date",  scope='world',  symbol='vaccines',  template=
"plotly_dark", title='Vaccination Count')
fig1.layout.updatemenus[0].buttons[0].args[1]["frame"]["duration"] = 400
fig1.update_geos(landcolor="red",oceancolor="#006994",showocean=True,
lakecolor="Blue"
)
fig1.update_traces(marker_coloraxis=None
)
fig1.show()
```

Number of vaccination available in the world the result obtained is shown in Fig. 4.

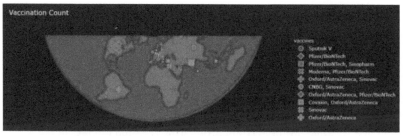

Fig. 4. Various vaccination count available round the world.

It is easily seen that the influence of Pfizer/Biotech in the USA is immense and it tends to grow with time. We can also see CNBG/strong Cinovac's influence beginning in China. Sinovac takes a sudden roll on the 14th in Turkey and starts to develop afterwards.

Step 4: Analysis of Vaccine Distributed the Most

```
s=data.drop_duplicates(subset=['iso_code'])['vaccines'].apply(lambdax:
x.split(',')) dic={}
for i in s:
for j in i :
if j[0]=='':
        k= j[1:]
elifj[-1]=='':
        k=j[:-1]
```

```
else:
        k=j
if k not in dic : dic[k]=1
else:
dic[k]+=1px.bar(x=list(dic.keys()),y=list(dic.values()),color=list(dic.
keys()),template='plotly_white',labels={'x':'VaccineName','y':'Total
Count'})
```

It is clearly observed that Pfizer in January 2021, followed by Moderna and Sinovacac, became the most common vaccine in the world.

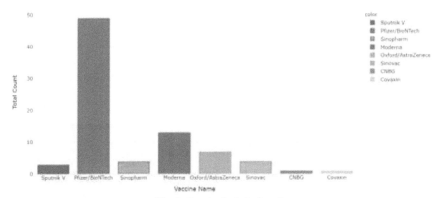

Fig. 5. Result obtain in Step 4.

Step 5: Observe the Deathrate after Vaccine

```
arr=[] index1=data.groupby(['country','date']).count().index index2=data2.
groupby(['Country','Date']).count().index for i in index2:
if i in index1: arr.append(1) else: arr.append(0)
data2['Vaccine_is_there']=arr
data2.head()
```

After a line indicated in the figures the vaccine was introduced in the country. The Country names are indicated on the figure for a better understanding of trends.

A variety of organizations are responsible for all the vaccines delivered globally and many are being currently created while some are being phased out and some are waiting for clearance.

Many nations, such as the UK, have chosen to use multiple vaccines to provide the people with the required ones.

Out[14]:

	Confirmed	Recovered	Deaths	Vaccine_is_there
count	7.180800e+04	7.180800e+04	71808.000000	71808.000000
mean	1.440877e+05	8.692296e+04	3921.234305	0.020221
std	8.780563e+05	5.053113e+05	19220.500408	0.140755
min	0.000000e+00	0.000000e+00	0.000000	0.000000
25%	6.800000e+01	1.600000e+01	0.000000	0.000000
50%	2.645500e+03	1.290000e+03	53.000000	0.000000
75%	3.695150e+04	1.803075e+04	638.000000	0.000000
max	2.592928e+07	1.040915e+07	436678.000000	1.000000

Fig. 6. Result obtain in Step 5.

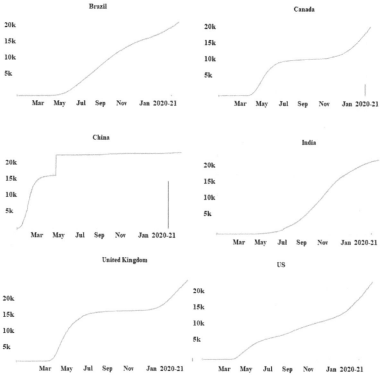

Fig. 7. Results of different countries showing effect on deathrate after vaccine.

However, where are the vaccines used? It is found that

- The vaccines developed by Oxford/AstraZeneca and Pfizer/BioNTech are the most commonly distributed around the world.
- More vaccines tend to be nearby and certain vaccines appear to protect landless or countries in the vicinity (i.e., Oxford/AstraZeneca, by itself for most of the African continent).
- In the third world countries the Oxford/AstraZeneca vaccine is more common. This makes sense because it is cost-free, so the countries with a lower GDP will get the vaccine.
- The Covaxin vaccine is actually being phased out but being used only in India, indicating that it has not yet been licensed for use in other countries.
- China (Sinopharm/Beijing, Sinopharm/Wuhan, Sinovac) and Russia (EpiVacCorona, Sputnik V) are the only countries that do not use vaccines produced outside their borders.
- According to the result, the United Arab Emirates and Hungary are currently rolling out the most vaccines, each coming from five different suppliers.

COVID-19 vaccines have been in clinical trials since mid-late 2020, and were formally rolled out for use in December 2020. They have since been adopted in more countries.

At this time, several nations are in the middle of their vaccine campaigns. Although several countries are yet to begin vaccination campaigns, others are

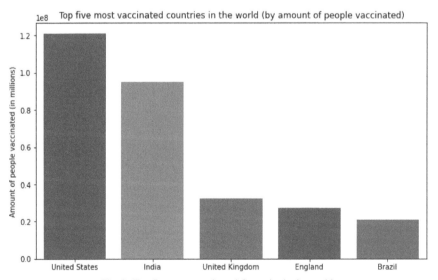

Fig. 8. Top five most vaccinated Countries in the world.

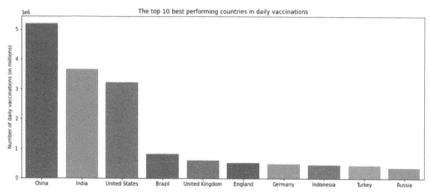

Fig. 9. Top ten best performing countries in daily vaccinations.

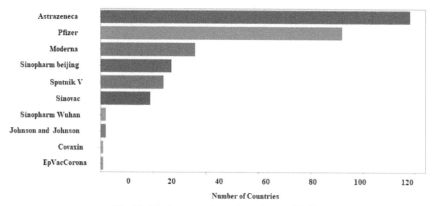

Fig. 10. Most commonly used vaccine worldwide.

already on their way to vaccinating their people. Which nations, on the other hand, have the highest and most successful vaccine programmes? The answer to the above question can be represented in Figs. 9–11.

The graph above depicts a very positive image of how vaccine programmes are improving around the world; not only is every country vaccinating more than their death tolls per day, but certain countries have reached significant margins for those that are vaccinated and their death tolls. India (with a massive gap margin of 173.4 percent *), the United States (138.8 percent *), Brazil (56 percent *), and Mexico (45.2 percent *) happen to be the highest performers in the top ten.

We've seen the vaccination numbers for each country, and they rise with population growth. However, what if we analyse from a different perspective, irrespective of population numbers? Are recovery rates related to the success

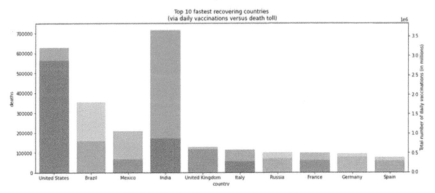

Fig. 11. Top ten fastest recovering countries.

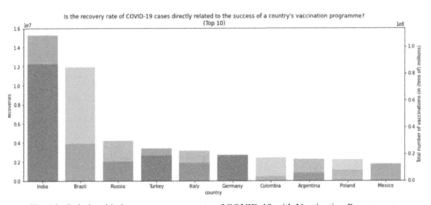

Fig. 12. Relationship between recovery rate of COVID-19 with Vaccination Programme.

of a vaccination program, and is that dependent on what kind of country it is economically (i.e., first, third-world, developing)?

The answer to this question can be understood from Fig. 12 given below.

4. Conclusion

In the top ten, we see some newcomers. India, Brazil, Italy, and Mexico all featuring on both maps, indicating that recovery rates are affecting the vaccination programme. The most intriguing aspect of this graph is that the wealthier countries (the United States, the United Kingdom, and so on) are not included in the top ten, implying that these countries were the worst hit and had poorer recovery rates (which makes sense, their death rates were higher). The results for Turkey and Germany are also interesting because the margins between recoveries and overall vaccinations are tiny. May be their populations more compliant with lockout limits, resulting in less infections.

"Vaccination intent is highest in China", according to the survey, where 80 percent of respondents strongly or somewhat agreed with the statement "If a COVID-19 vaccine were available", I would get it. Countries with a reasonably high degree of intent include Brazil (78%), the United Kingdom (77%), Mexico (77%), Australia (75%) and South Korea (75%). Among the countries surveyed, South Africa (53 percent), Russia (43 percent) and France were those whose populations recorded the lowest intent (40%). It is clearly observed from Fig. 7 that the deathrate is either stagnant or low in almost all countries but it is still severe in US and UK where a new mutant of COVID 19 has been found.

5. Future Scope

This research study will be very helpful for the researchers working in the same field.

References

[1] Singh, R. and Bhatia, A. 2018. Sentiment analysis using Machine Learning technique to predict outbreaks and epidemics. Int. J. Adv. Sci. Res. 3(2): 19–24.

[2] Ting, Daniel Shu Wei, Lawrence Carin, Victor Dzau, Tien Y. Wong et al. 2020. Digital technology and COVID-19. Nature Medicine (2020): 1–3.

[3] Benvenuto, Domenico, Marta Giovanetti, Lazzaro Vassallo, Silvia Angeletti et al. 2020. Application the ARIMA model on the COVID-2019 epidemic dataset. Data in Brief (2020): 105340.

[4] Deb, Soudeep and Manidipa Majumdar. 2020. A time series method to analyze incidence pattern and estimate reproduction number of COVID-19. arXiv preprint arXiv: 2003.10655.

[5] Kucharski, Adam J., Timothy W. Russell, Charlie Diamond et al. 2020. Early dynamics of transmission and control of COVID-19: a mathematical modelling study. The Lancet Infectious Diseases.

[6] Dey, Samrat Kumar, Md Mahbubur Rahman, Umme Raihan Siddiqi, Arpita Howlader et al. 2020. Analyzing the epidemiological outbreak of COVID-19: A visual exploratory data analysis (EDA) approach. Journal of Medical Virology.

[7] Meredith, R., Andrew S. Azman, Nicholas G. Reich and Justin Lessler. 2020. The incubation period of coronavirus disease 2019 (COVID-19) from publicly reported confirmed cases: estimation and application. Annals of Internal Medicine.

[8] Singer, H. M. 2020. Short-term predictions of country-specific COVID-19 infection rates based on power law scaling exponents. arXiv preprint arXiv: 2003.11997.

[9] Awad, Mariette and Rahul Khanna. 2015. Support vector regression. pp. 67–80. *In*: Efficient Learning Machines. A press, Berkeley, CA.

[10] De Castro, Yohann, Fabrice Gamboa, Didier Henrion, Roxana Hess et al. 2019. Approximate optimal designs for multivariate polynomial regression. The Annals of Statistics 47(1): 127–155.

[11] Greff, Klaus, Rupesh K. Srivastava, JanKoutn´ık, Bas R. Steunebrink and Ju¨rgen Schmidhuber. 2016. LSTM: A search space odyssey. IEEE Transactions on Neural Networks and Learning Systems 28(10): 2222–2232.

[12] Ojha, U. and Goel, S. 2017. A study on prediction of breast cancer recurrence using data mining techniques. Proc. 7th Int. Conf. Conflu. 2017 Cloud Comput. Data Sci. Eng. pp. 527–530.

[13] Ghosh, S., Mondal, S. and Ghosh, B. 2014. A comparative study of breast cancer detection based on SVM and MLP BPN classifier. 1st Int. Conf. Autom. Control. Energy Syst. - 2014, ACES 2014. pp. 1–4.

[14] Osareh, A. and Shadgar, B. 2010. Machine learning techniques to diagnose breast cancer. 2010 5th Int. Symp. Heal. Informatics Bioinformatics, HIBIT 2010, pp. 114–120.

[15] Bazazeh, D. and Shubair, R. 2017. Comparative study of machine learning algorithms for breast cancer detection and diagnosis. Int. Conf. Electron. Devices, Syst. Appl., pp. 2–5.

[16] Azmi, M. S. B. M. and Cob, Z. C. 2010. Breast cancer prediction based on backpropagation algorithm. Proceeding, 2010 IEEE Student Conf. Res. Dev. - Eng. Innov. Beyond, SCOReD 2010, no. SCOReD, pp. 164–168.

[17] Gayathri, B. M. and Sumathi, C. P. 2017. Comparative study of relevance vector machine with various machine learning techniques used for detecting breast cancer. 2016 IEEE Int. Conf. Comput. Intell. Comput. Res. ICCIC 2016, pp. 0–4.

[18] Purva Agarwal and Pawan Whig. 2017. Low delay based 4 Bit QSD adder/subtraction number system by reversible logic gate. 2016 8th International Conference on Computational Intelligence and Communication Networks (CICN), IEEE Xplore: 26 October 2017.

[19] Jacob B. Chacko and Pawan Whig. 2017. Low delay based full adder/subtractor by MIG and COG reversible logic gate. 2016 8th International Conference on Computational Intelligence and Communication Networks (CICN), IEEE Xplore: 26 October 2017.

[20] Ajay Rupani, Pawan Whig, Gajendra Sujediya and Piyush Vyas. 2017. A robust technique for image processing based on interfacing of Raspberry-Pi and FPGA using IoT. International Conference on Computer, Communications and Electronics (Comptelix), IEEE Xplore: 18 August 2017.

[21] Pawan Whig and Ahmad, S. N. 2014. Simulation of linear dynamic macro model of photo catalytic sensor in SPICE. Compel, The International Journal of Computation and Mathematics in Electrical and Electronic Engineering, Vol. 33, No. 1/2. ISSN: 0332-1649 (SCI, ISI index).

[22] Vaibhav Bhatia and Pawan Whig. 2013. A secured dual tune multi frequency based smart elevator control system. International Journal of Research in Engineering and Advanced Technology Vol. 4 Issue 1. ISSN (Online): 2319–1163.

[23] Pawan Whig and Ahmad, S. N. 2013. A novel pseudo NMOS integrated ISFET device for water quality monitoring. Active and Passive Components Hindawi Article i.d 258970. Vol. 1 Issue 1(Scopus). ISSN 0882-7516.

[24] Vaibhav Bhatia and Pawan Whig. 2013. Modeling and simulation of electrical load control system using RF technology. International Journal of Multidisplinary Science and Engineering 4(2): 44–47. ISSN 2045-7057.

[25] Pawan Whig and Ahmad, S. N. 2014. Development of economical ASIC for PCS for water quality monitoring. Journal of Circuit System and Computers 23(6): 1–13. ISSN: 0218-1266 (SCI, ISI index).

[26] Aastha Sharma, Abhishek Kumar and Pawan Whig. 2015. On the performance of CDTA based novel analog inverse low pass filter using 0.35 μm CMOS parameter. International Journal of Science, Technology & Management 4(1): 594–601. ISSN No: 1460-6720.

[27] Pawan Whig and Ahmad, S. N. 2016. Simulation and performance analysis of low power quasi floating gate PCS model. International Journal of Intelligent Engineering and Systems 9(2): 8–13 (Scopus). ISSN: 2185-3118.

[28] Pawan Whig and Ahmad, S. N. 2016. Ultraviolet photo catalytic oxidation (UVPCO) sensor for air and surface sanitizers using CS amplifier. Global Journal of Researches in Engineering: F, 16(6): 1–13. ISSN Numbers: Online: 2249-4596. Print: 0975-5861 DOI: 10.17406/GJRE.

[29] Rashmi Sinha, Shweeta Prashar and Pawan Whig. 2015. Effect of output error on fuzzy interface for VDRC of second order systems. International Journal of Computer Applications 125(13): 2015. ISSN: 0975–8887.

CHAPTER 3
Swarm Intelligence and Machine Learning Algorithms for Cancer Diagnosis

Pankaj Sharma,[1,]* *Vinay Jain*[2] and *Mukul Tailang*[3]

1. Introduction

Artificial intelligence (AI) is a field of study that uses programs to mimic human intelligence [1]. Artificial intelligence does have the capacity to alter contemporary technology and cultural socioeconomic processes [2]. Machine learning (ML) is a part of artificial intelligence (AI) that uses numerical and computational analysis to improve computer effectiveness (Fig. 1). Despite being properly taught, machine learning algorithms (MLAs) construct a predictive method based on existing data, alluded to as "training set," to produce forecasts or inferences. Deep learning is a class of machine learning that uses number of layered convolutional neural networks to comprehend data (Fig. 1).

The term "deep learning" is used in various domains to describe a group of current techniques that, when integrated, have achieved considerable efficiency gains over the best existing machine learning algorithms [3]. In 2018, approximately 18 million extra cancer patients (known as recurrence) were recognized internationally, according to data. Among some of the

[1] Department of Pharmaceutics, ShriRam College of Pharmacy, Banmore, Morena (M.P.)-476444, India.
[2] Department of Pharmacognosy, ShriRam College of Pharmacy, Banmore, Morena (M.P.)-476444, India.
[3] School of Studies in Pharmaceutical Sciences, Jiwaji University, Gwalior (M.P.)-474011, India.
Emails: vinni77@gmail.com; mukultailang@gmail.com
* Corresponding author: pankajsharma223@gmail.com

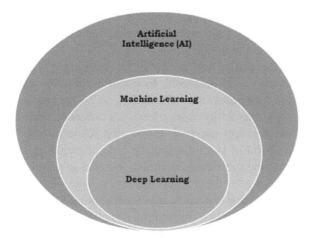

Fig. 1. Linkage with machine learning, deep learning, and artificial intelligence.

18 million cases, 5 million cases of mammary, pharyngeal, ovarian, and bowel cancer could have been eradicated or managed more successfully if detected earlier. According to some of the science based studies conducted in various nations, early detection, appropriate vetting, and definitive diagnosis of the extent and type of carcinoma may very well have markedly enhanced carcinoma patient life expectancies, quicker support, and a better quality of life as sick people, along with substantial price reduction and sophistication of carcinoma hospital facilities. If early cancer detection was handled by public health services, the commercial healthcare industry, and cancer sufferers, the financial strain of cancer would have just been reduced and not only in certain countries, but internationally as well [4–5]. Cancer is the most common cause of death and occurrence globally, responsible for nearly one-fifth among all related deaths, including 2.1 million new cases as well as 1.8 million deaths expected in 2018. In urbanization, cigarettes and poor air quality are the leading causes of lung cancer [6–7].

Machine learning is not a new notion in cancer research. Decision trees (DTs) and convolutional neural networks have been utilised in cancer diagnosis and monitoring for almost two decades [8–10]. MLAs are currently utilised in a wide range of activities, from detecting and describing tumors combining CRT and X-ray images [11–12] to define malignancies utilising proteomic and genomic (microarray) assessment [13–15]. Approximately 1500 publications on oncology using machine learning have indeed been released, including the most comprehensive PubMed data. However, the vast majority of these papers are concerned with applying machine learning techniques to detect, identify, track, or distinguish tumors and certain other malignancies. To look at it another way, machine learning has mostly been applied to cancer detection and prevention [16]. Professionals in cancer diagnosis and prognosis have recently attempted to use machine learning for cancer diagnosis and prediction.

Carcinoma forecasting and prognosis have different fundamental goals than cancer detection and diagnosis. In tumor progression and forecasting, three predictive attributes are considered: (1) carcinoma vulnerability forecasting (i.e., vulnerability assessment); (2) cancer recurrence likelihood; and (3) predicting cancer survival. In the first scenario, one is seeking to predict the likelihood of getting a specific type of cancer prior to the onset of the sickness. In the subsequent situation, one tries to predict the likelihood of cancer recurrence after the sickness appears to have passed. Considering a cancer diagnosis, the third scenario entails attempting to predict a prognosis (life expectancy, survivability, tumor-drug sensitivity, growth). The efficacy of the prognostic forecast is impacted by the correctness or constancy of the identification in the latter two cases. A clinical prognosis, but on the other extreme, can only be given following a clinical diagnosis, and a prospective projection must take into account more than the diagnosis itself [17].

The use of machine learning and data mining for screening methods is now becoming extremely prevalent, and cancer diagnosis has been one of the chronic ailments wherein the categorization problem is critical. As a consequence, machine learning algorithms can help clinicians make an accurate cancer diagnosis as well as accurately identify a tumor as malignant or benign. The examination of health information and the judgement of physicians and specialists is without a doubt the most important factor in identification, although specialist systems and AI tools, as well as ML for categorization tasks, continue to substantially aid clinicians and professionals.

2. Machine Learning Approaches

Prior to actually delving into whatever machine learning approaches are best for specific circumstances, it's important to understand exactly what machine learning is and isn't. ML is a branch of AI study that employs a blend of mathematics, computational, and probabilistic techniques to "grasp" information from past occurrences and afterwards applies this information to discover new information and trends, and anticipate future trends [18]. Machine learning is frequently treated in a similar manner that statistics is to analyse and display information. Machine learning techniques, despite statistics, could use fundamental preconditions (IF, ELSE, THEN), Boolean logic (AND, NOT, OR), dependent possibilities (the likelihood of X given Y), and unconventional optimization approaches to model information or discern similarities. These latter methods are comparable to how people learn and recognise information.

Supervised and unsupervised learning and reinforcement learning are the three types of machine learning processes. The set of statistics input into the system in supervised learning is connected to a preset outcome. These tasks are generally classified as being either regression or categorisations, such as estimating the chance that somehow a malignancy will progress to a specific care [19]. The information is separated into two cohorts: development and examination. The earlier is being used to construct a mathematical formula, although the latter is included to assess its universal applicability. Unsupervised learning is useful whenever a specific outcome is uncertain and whenever investigators are looking for new tendencies in data [20–21]. A knowledge base assists in assessing how informative or intriguing unsupervised performance of the model is. Additional data testing will be performed for both unsupervised and supervised techniques to be assessed for generalizability [20]. Reinforcement learning is a form of learning which emphasizes on training a machine to govern itself in order to achieve a long-term objective by optimizing a quantitative outcome measure. The individuals receive only incomplete information on their expectations in reinforcement learning, as opposed to assisted grasping. Additionally, predictions might have protracted repercussions by influencing the future status of the controlled system. As a consequence, time is extremely important. The goal of reinforcement learning is to develop efficient learning algorithms whilst fully comprehending their advantages and disadvantages [21].

The prominent machine learning algorithms utilised from the above forms of machine learning are covered in this section.

2.1 Bayesian Network (BNs)

Bayesian Networks are pictorial constructions that permit an undetermined subject to be represented and reasoned about. BNs are a mix of probability theory and pattern recognition. By itemising a collection of conditional suppositions of independence and also a collection of likelihood function; BNs describe the probabilistic model regulating a number of indicators. The network's network comprises discrete or continuous parameters, with arcs illustrating their interdependencies. In Bayesian Networks [22], conditional independence is a crucial concept.

Given the likelihoods of C, B, and the organization in Fig. 2, the possibility of J is independent and identically distributed on A, as shown in the following equation:

$$P(J \mid A, C, B) = P(J \mid C, B)$$

Each vertex is deemed conditionally independent of quasi if its quick predecessors are present. As a corollary, understanding the target characteristic becomes less challenging.

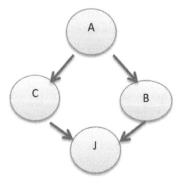

Fig. 2. The framework comprises of a Bayesian Network.

2.2 K-Nearest Neighbors (KNNs)

It is one of the most effective algorithms and techniques for regression and categorization issues. To differentiate new facts from news feed locations, KNN algorithms mostly utilise resemblance checks. The area is classified by a plurality vote among its neighbors [23].

2.3 Naïve Bayes (NBs)

It focuses on probabilistic approaches that contrast one feature of a category with something else that has distinct beliefs and attributes which may be completely different from each other. The category with the greatest chance is thought to become the most perfect [24]. Naive Bayes may well be conceived of as a particular case of Bayesian Networks (Fig. 3).

Every parameter (X1) in Fig. 3 is an independent event of many other elements if it belongs to the same category (Z). The Naive Bayes presupposition, on either extreme, seems to be more restrictive than that of the Bayesian Network's worldwide supposition, which governs the parameters by establishing a set of conditional independence assumptions between the features as well as a collection of likelihood functions.

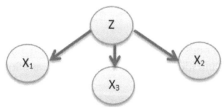

Fig. 3. Naive Bayes a unique situation.

2.4 Random Forest (RF)

It's used in the construction of numerous decision trees. Random Forest chooses the ultimate result based on the number of leaves [25].

Figure 4 is a representation of the tree-like structure of DTs. Every variable (A, B, C) is represented by a circle, and the choice results are represented by squares (Class 1, Class 2). T(1–3) represents the limits (classification rules) necessary to successfully categorise each parameter to a class mark.

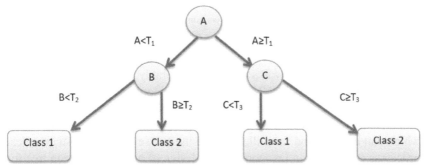

Fig. 4. Algorithms for Random Forests.

2.5 Convolutional Neural Networks (CNNs)

A type of thinking based on the human brain is known as a convolutional neural network (CNN) [26]. ANNs have become a common research topic in recent decades, with a growing number of researchers using them. CNNs also yielded a slew of milestones, including significant advancements in BC diagnosis and classification in the very initial stages [27]. A CNN model usually consists of three layers: input, secret, and output (Fig. 5) [28]. Layers are made up of interacting neurons that have a nonlinear transformation activation mechanism that helps to improve nonlinear expression. The data is received by the input layer and then sent to the secret layer, which processes it and sends the learning outcomes to the output layer. The classification effects appear in the output layer. Determined by the issues, though, the method of instructions and CNNs can entail long pivotal chains of computational phases. Back propagation (BP), a precious structured gradient fall algorithm, has been widely used since 1986, particularly for medical data [29]. Despite the fact that the BP algorithm is used, it has several flaws when dealing with large amounts of data. Since BP calculations are lengthy, preparation takes a long time; as a result, a refined BP algorithm is seldom applied in functional utilization.

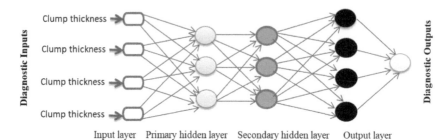

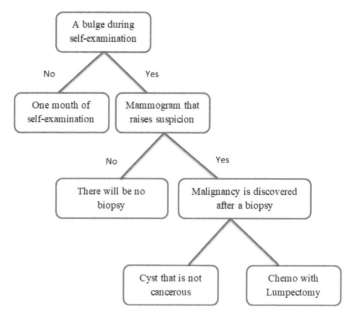

Input layer   Primary hidden layer   Secondary hidden layer   Output layer

Fig. 5. A small explanation of how a CNN is trained to predict diagnostic findings using four inputs and 2 hidden layers using 4 neurons.

2.6 Decision Tree (DT$_s$)

Convolutional neural networks have a considerably more complicated logic than decision trees (DTs) (CNNs). A decision tree topology is a special flowing chart or network of decisions (nodes) and probable consequences (divisions or leaves) that is used to construct a goal-setting strategy [18–30]. Decision trees, which had been around for millennia, are used in several diagnostic medical processes (especially in taxonomy). The design of a simple decision tree for identifying breast cancer is shown in Fig. 6. Usually, decision tree algorithms are developed with the assistance of experts and fine-tuned over period, or they are altered to fit resource restrictions or minimise risk. However, there

Fig. 6. A set of fundamental decision trees for the management and therapy of cancer [32].

are decision tree learners who can build decision trees automatically given a labelled collection of training data. When using decision tree trainers to categorise the data, the tree's leaves define classifications, and the branches define function combinations that contribute to certain classifications. A decision tree is created by gradually dividing the named training data into subdivided sets depending on a logical or numerical examination [30]. This method is replicated recursively on each dependent subdivided set until no more separation is feasible or a unique classification is obtained. Decision trees have several pluses: they are easy to understand and scrutinise, they require little data scheduling, they can handle a wide variety of data types, which include present value (named), computation, and categorical variables, they focus on providing rigorous classifiers, those that are quick to "peruse," and they can be authenticated using a normality test. DTs, on the other hand, often do not react as well as CNNs in more complex categorization tasks [31].

2.7 Support Vector Machine (SVMs)

SVMs are well-known in the world of machine learning, yet they are practically unknown in the field of oncology diagnosis and prognosis. A graphical representation of dots, including primary tumor vs the number of axillary nodules (for breast cancer) in individuals with excellent and terrible prognostications, is the simplest method to explain exactly how SVMs work (Fig. 7). There are two separate groups of people. The SVMs machine learner would look for a calculation for a path which would split the two clusters more successfully. If additional factors were displayed, the separation line would have become a plane (say volume, metastases, and estrogen receptor extent). If there were more parameters, the separation would be described by a hyperplane. The hyperplane is defined by the assistance vectors, which are

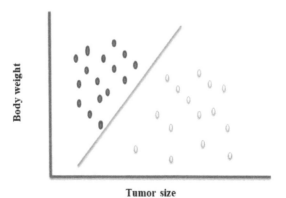

Tumor size

Fig. 7. Algorithm for calculating the machine's support vector. Tumors are classified characteristics such as patient size and mass.

a subgroup of the vertices from of the 2 categories. Colloquially, the SVMs algorithm creates a hyperplane which splits the information into two groups with the largest margin—that really is, the distance here between hyperplane and the closest instances (the margin) is greatest. SVMs could also be used to perform quasi categorization using a non-linear kernel. A quasi kernel is a computational method for converting data from a particular linear set to a particular non-linear set. Applying various kernels on multiple datasets may substantially enhance the performance of a SVMs classifier [33].

2.8 Dataset

The information utilised seems to be from Wisconsin, so it enables you to get a brief summary of something like the database's components so we may use them as an information provider. The information, which will be divided into 32 columns and can therefore assist us in predicting whether another cancer issue is malignant or benign, as well as aid patients and families by choosing the ideal specificity and getting the desired effects, was created by Dr. William H. Wolberg [34]. So that they don't yet impact individuals who have the same irregular growth and possibly proceeding to manage them by attempting prior report investigations resulting in an improved result each time by conducting close supervision using algorithms on them can facilitate getting specificity so that they can start serving as an inkling of which algorithm is preferable due to a recognizable distinction and an ability to give additional precision.

3. Machine Learning Applications in Medical Practice

For generations, machine learning has been utilised in a variety of approaches. Throughout the last generation, the use of machine learning application in medical practice began to improve Treatment performance by speeding up processes and boosting reliability, leading the path for better overall healthcare. MLAs can help with acute management and diagnosis while also enhancing physicians' abilities. In principle, an algorithm requires any source and uses mathematics and reasoning to generate an output. Additionally, when provided with fresh intakes, a machine learning algorithm employs both inputs and outputs simultaneously to "understand" the data and generate results. Machine learning is a technique for teaching machines to understand from their experiences. New advancements in Artificial Intelligence (AI) are set to revolutionize and transform medical procedure, with numerous intelligent automations in the diagnosis, management, and forecasting of health outcomes. Most clinicians spent far more time on recording input and paperwork than actually chatting with individuals through office consultations [35–36]. The analytical skills of extremely developed programs have been tested and compared to traditional diagnostic tools and professionals, and they have proven to be extremely effective in the identification and accurate prognosis of a range of diseases [37–38].

3.1 Machine Learning Approach for Prostate Cancer Diagnosis

In 1991, Schnall et al. [39] sought to combine pathology and radiography, but only had little success in terms of pathologic and radiographic features. In 2012, Ward et al. [40], used an image-directed cutting approach based on strand-shaped ideal framework identifiers to streamline image capture from MRI histopathology and *in vivo*. Since 2014, Litjens et al. [41] have been utilising several frameworks (evaluating algorithms of prostate pictures, photons for easy image recognition) to partition and characterise MRI images, and their effort forms the foundation for even more subsequently identified and used techniques.

Machine intelligence and CNN techniques show potential for prostate cancer detection and prognosis forecasting. There is a tremendous need for more study because there is very little information. There is a tremendous need for more study due to shortage of information. Diagnostic imaging, genetics, histopathological, and treatment each have the potential to enhance disease atomisation and, as a corollary, improving clinical treatment in the hospital. In the near future, the possibility of learning algorithms in prostate surgical intervention will improve planning and operation outcomes, both in terms of increasing patient outcomes and in terms of teaching and assessing surgical abilities.

3.2 Breast Cancer Detection and Prognosis using Machine Learning

The Wisconsin breast cancer diagnosis (WBCD) dataset was first released in the early 1990s, and throughout the following 3 decades, a variety of machine learning methods were used to assess it. CNNs, SVMs, and DTs, along with their usage in breast cancer detection and prognosis, are all analysed using the WBCD [42–43]. The 4 procedures are listed in detail in Table 1 at the conclusion of this chapter, containing algorithm descriptions, citations, categorization preciseness, and specimen approaches. Table 1 includes a complete description of the 4 stages, algorithm descriptions, citations, sample procedures, and classification accuracy. These algorithms can help in the creation of medical services and can act as a second opinion for clinicians when they create a definitive choice, improving the reliability of their decisions. The Machine learning methods detailed in the previous subparts are all used to assess free software benchmark functions, commencing with SVMs, CNNs, DTs, and k-NNs.

Table 1. In WBCD, a concise summary of ML approaches including the use of DTs, CNNs, SVMs, and k-NNs is provided [53].

Citations	Methods	Specimen Strategies	Categorization Preciseness (%)
Quinlan 1996 [44]	DT	Cross-validation ten times	94.74
Bennett and Blue 1998 [45]	SVM	Cross-validation five times	97.20
Setiono 2000 [46]	CNN	Cross-validation ten times	97.97
Sarkar and Leong 2000 [47]	k-NN	Training-testing ratio of 50/50	98.25
Pach and Abonyi 2006 [48]	DT	Cross-validation ten times	95.27
Karabatak and Ince 2009 [49]	CNN	Cross-validation three times	97.40
Chen et al. 2011 [50]	SVM	Training-testing ratio 20/80	96.87
Koyuncu and Ceylan 2013 [51]	CNN	Training-testing ratio of 50/50	98.05
Kumar et al. 2017 [52]	SVM-Naive Bayes	Cross-validation ten times	97.13

3.3 Lung Cancer Detection and Prognosis Uses of Machine Learning

To test if they might identify lung cancer early and accurately, a lot of scientists looked at samples from malignant (cancerous malignancy) and quasi cell lung carcinoma. The study's goal was to determine whether ML algorithms were most efficient in identifying lung cancer. They employed small cell lung cancer and quasi cellular lung carcinoma labelled biopsy tissues. The specimen images were diagnosed by two expert physicians (over 20 years). Several deep learning techniques also were able to differentiate among quasi and cancer specimens. According to the findings of the research, machine learning algorithms have a region underneath the curve reliability (AUC) of 0.8810 to 0.9119. According to scientists, machine learning assessment might aid in the identification of lung cancer samples whilst retaining a detection performance comparable to that of a human expert observer [54].

3.4 Uses of Machine Learning in the Detection and Prognosis of Mouth Cancer

As per PubMed data, the bulk of publications released on ML and cancer were linked with applying specific learning techniques to distinguish, detect, monitor, or distinguish tumors as well as other abnormalities. For a long time, experts have struggled to anticipate the conclusion of chemotherapeutic agents. Throughout the last 2 decades, a body of knowledge about ML and malignancy prognosis has grown. The majority of these trials employ gene expression profiles, clinical variables, and histological characteristics as variables to the prognostic procedure [55]. In their study, Chang et al. integrated feature options and machine learning techniques to predict mouth cancer outcomes. As per their findings [56], including both genetic and clinicopathologic indicators enhanced survival. In another assessment, Exarches et al. set out to categorise the factors that impact the establishment of oral squamous cell carcinoma that, as both a consequence, predict future recurrence of similar illnesses. They presented a multi-parametric advanced analytical technique based on a range of data sources, such as medical genetic and imaging information. This study described how machine learning classifiers may be utilised to incorporate diverse data sources and produce accurate findings in tumor recurrence prediction [57].

3.5 Skin Cancer Detection and Prognosis using Machine Learning

There are numerous intriguing opportunities regarding machine learning in the dermatologist's clinic. Along with its capacity to make skin cancer screenings more visible and specialists' processes more efficient, CNN categorization of images has received the greatest attention. Non-melanoma skin cancer (NMSC), atopic dermatitis, melanoma, rosacea, psoriasis, and oncomycosis are the most common skin lesions studied in dermatological research that utilise machine learning to identify skin diseases. CNN is utilised extensively in these investigations for picture segmentation and retrieval. Originally, a pre-trained model CNN (AlexNet) was utilised solely to collect characteristics, which has been subsequently categorized as another more basic machine learning method such as SVMs 58–60 or k-nearest neighbors. The bulk of CNNs will presently use end-to-end examination to extract elements and recognize pictures. Brinker and coworkers reported that CNNs had higher reliability and selectivity in melanoma categorization than board-validated dermatologists in the college [61–62].

3.6 Colorectal Cancer Detection and Prognosis using Machine Learning

A diagnosis, which would be founded on the context of multi data gathering and clinical practice, is among the most essential ideas in healthcare. Due to the obvious wide range of tumor symptoms, pace of malignancy development, personality differences, and medication susceptibility, it's challenging to make an appropriate tumor assessment. Doctors may be able to utilise machine learning to aid in the subjective identification and treatment of colorectal cancer, which is currently limited to pathological biopsies and colonoscopy [63]. Machine learning systems can detect and interpret tissue sample remotely, substantially improving diagnostic accuracy whilst reducing time and price [64]. Rathore et al. [65] developed a novel bowel cancer diagnostic (BCD) method based on the SVM radial basis work algorithm, which identifies benign and diseased colon biopsies pictures and assigns malignant scores automatically. When compared to conventional techniques, our BCD technique provides better cancer detection (precision 95.40%) and rating (accuracy 93.47%) effectiveness. An identical group then developed a hybrid character-space-based colonic categorization (HCS-CC) approach based on just this technique, which uses geometry, geometric characteristics, and texture to describe biopsy specimen pictures [66]. By utilizing a SVM as a filtering approach to determine 176 individuals, the HCS-CC approach achieved an accuracy result of 98.07%. By integrating a sub-patch value color histogram with generalized least SVM, Yang et al. [67] developed a novel machine learning method for bowel cancer histopathology. This method not just displays the color and geographical complexity of tumor pictures, but also displays contradictory information by providing excellent tumor classification performance (96.78 percent).

4. Conclusion

By offering both healthcare practitioners and individuals exposure to massive data, well-being digitalization is altering clinical workflow. Experience-based healthcare is rapidly being displaced by scientific proof and patient-centered treatment. Machine learning technology will have a big influence on the match of cancer in the not-too-distant future. Both doctors and researchers have to be ready for the coming rebellion [68]. Major innovations have come to the aid during last decade using machine learning techniques that have the ability to revolutionize and transform medical screening processes with various sophisticated technologies in the diagnosis, treatment, and forecasting of outcomes for cancer patients. Machine learning-assisted timely screening has advanced substantially in recent years, leading to greater specificity, cancer identification, and fewer false positive rates. Data collection, medical tests, and diagnostic evaluations for individuals will benefit from machine learning. Furthermore, machine learning techniques may read and evaluate obtained information and healthcare results, provide recommendations based on the obtained wellness data and matching it to past expertise and experience, and provide therapy. Ultimately, machine learning will utilize assessments to choose appropriate recovery choices for patients while re-evaluating the diagnostics and therapeutics. While AI could be preferable to doctors in some situations, the usage of machine learning might raise ethical concerns and generate a host of other issues. The overwhelming majority of AI applications in initial cancer diagnosis were beneficial to the healthcare professionals, and like machine learning developed with contemporary algorithms, so did their effectiveness. According to scientists and professionals in the field of tumor scanner image analysis, machine learning techniques can assist in reducing diagnostic time, expenses, and patient uncertainties even while delivering performance in vital operations for most cancer kinds. Machine learning techniques are now being used to help medical fields that rely on image information, such as histopathology, radiography, and cancer scare. Machine learning specializes in recognizing dynamic trends in visual information and can give a thorough analysis autonomously. As machine learning methods are applied into the medical workflow as a means to aid clinicians and the medical system, highly consistent and accurate radiological tests may be done.

5. Funding

There was no funding for this project from any source.

6. Acknowledgments

Suman Jain (Director, School of Studies in Pharmaceutical Sciences, Jiwaji University, Gwalior) and Navneet Garud were especially helpful to the writers (Department of Pharmaceutics, School of studies in pharmaceutical sciences, Jiwaji University, Gwalior).

References

[1] Rajkomar, A., Dean, J. and Kohane, I. 2019. Machine learning in medicine. N. Engl. J. Med. 380: 1347–1358.

[2] Geuna, A., Guerzoni, M., Nuccio, M., Pammolli, F. and Rungi, A. 2017. Digital disruption and the transformation of Italian manufacturing. Available at: https://www.aspeninstitute. it/aspeniaonline/article/digital-disruption-and-manufacturing-transformation-italian-case-study.

[3] LeCun, Y., Bengio, Y. and Hinton, G. 2015. Deep learning. Nature 521: 436–444.

[4] Hawkes, N. 2019. Cancer survival data emphasise importance of early diagnosis. Brit. Med. J. 364: 1408–1411.

[5] Cancer Research UK. Why is early diagnosis important? Finding and treating cancer at an early stage can save lives. Available at: [https://www.cancerresearchuk.org/about-cancer/cancer-symptoms/why-is-early-diagnosis-important].

[6] Thun, M. J., Henley, S. J. and Travis, W. D. 2018. Lung cancer. pp. 519–542. *In*: Thun, M. J., Linet, M. S., Cerhan, J. R., Haiman, C. A. and Schottenfeld, D. (eds.). Cancer Epidemiology and Prevention. 4th ed. New York, NY: Oxford University Press.

[7] Jha, P. 2009. Avoidable global cancer deaths and total deaths from smoking. Nat. Rev. Cancer 9: 655–664.

[8] Simes, R. J. 1985. Treatment selection for cancer patients: application of statistical decision theory to the treatment of advanced ovarian cancer. J. Chronic Dis. 38: 171–186.

[9] Maclin, P. S., Dempsey, J., Brooks, J. and Rand, J. 1991. Using neural networks to diagnose cancer. Journal of Medical Systems 15(1): 11–19.

[10] Cicchetti, D. V. 1992. Neural networks and diagnosis in the clinical laboratory: state of the art. Clin. Chem. 38: 9–10.

[11] Petricoin, E. F. and Liotta, L. A. 2004. SELDI-TOF-based serum proteomic pattern diagnostics for early detection of cancer. Curr. Opin. Biotechnol. 15: 24–30.

[12] Bocchi, L., Coppini, G., Nori, J. and Valli, G. 2004. Detection of single and clustered microcalcifications in mammograms using fractals models and neural networks. Med. Eng. Phys. 26: 303–312.

[13] Zhou, X., Liu, K. Y. and Wong, S. T. 2004. Cancer classification and prediction using logistic regression with Bayesian gene selection. J. Biomed. Inform. 37: 249–259.

[14] Dettling, M. 2004. BagBoosting for tumor classification with gene expression data. Bioinformatics 20: 3583–3593.

[15] Wang, J. X., Zhang, B., Yu, J. K., Liu, J., Yang, M. Q. et al. 2005. Application of serum protein fingerprinting coupled with artificial neural network model in diagnosis of hepatocellular carcinoma. Chinese Medical Journal 118(15): 1278–1284.

[16] Mccarthy, J. F., Marx, K. A., Hoffman, P. E., Gee, A. G., O'neil, P., Ujwal, M. L. et al. 2004. Applications of machine learning and high-dimensional visualization in cancer detection, diagnosis, and management. Ann. N Y Acad. Sci. 1020: 239–262.

[17] Hagerty, R. G., Butow, P. N., Ellis, P. M., Dimitry, S., Tattersall, M. H. et al. 2005. Communicating prognosis in cancer care: a systematic review of the literature. Ann. Oncol. 16: 1005–1053.

[18] Mitchell, T. 1997. Machine Learning. New York: McGraw Hill.

[19] Buch, V. H., Ahmed, I. and Maruthappu, M. 2018. Artificial intelligence in medicine: Current trends and future possibilities. Br. J. Gen. Pract. 68: 143–144.

[20] Liu, Y., Chen, P. H., Krause, J. and Peng, L. 2019. How to read articles that use machine learning: Users' guides to the medical literature. JAMA 322: 1806–1816.

[21] Csaba Szepesvari. 2010. Algorithms for Reinforcement Learning: Synthesis Lectures on Artificial Intelligence and Machine Learning. Morgan & Claypool Publishers, Williston. 4(1): 1–103 (https://doi.org/10.2200/S00268ED1V01Y201005AIM009).

[22] Adel Aloraini. 2012. Different machine learning algorithms for breast cancer diagnosis. (IJAIA) 3(6): 21.

[23] Sivakami, K. and Saraswathi, N. 2015. Mining big data: breast cancer prediction using DT-SVM hybrid model. IJSEAS 5: 418–429.

[24] Megha Rathi and Arun Kumar Singh. 2012. Breast cancer prediction using Naïve Bayes Classifie. Int. J. Inf. Technol. 1(2): 77–80.

[25] Cuong Nguyen, Yong Wang and Ha Nam Nguyen. 2013. Random forest classifier combined with feature selection for breast cancer diagnosis and prognostic. J. Biomed. Sci. and Eng. 6: 551–560.

[26] Hinton, G. E. 1992. How neural networks learn from experience. Sci. Am. 267: 144–151.

[27] Furundzic, D., Djordjevic, M. and Bekic, A. J. 1998. Neural networks approach to early breast cancer detection. J. Syst. Archit. 44: 617–633.

[28] Minsky, M. and Papert, S. 1969. Perceptrons. MIT Press: Cambridge, MA, USA.

[29] Rumelhart, D. E. and Mcclellend, J. L. 1986. Parallel Distributed Processing: Explorations in the Microstructure of Cognition. MIT Press: Cambridge, MA, USA.

[30] Quinlan, J. R. 1986. Induction of decision trees. Machine Learning 1: 81–106.

[31] Atlas, L. E., Cole, R. A., Connor, J. T., El-Sharkawi, M. A., Marks, II R. J. et al. 1990. Performance comparisons between backpropagation networks and classification trees on three real-world applications. NeurIPS 2: 622–629.

[32] Joseph, A. and Cruz David, S. W. 2006. Applications of machine learning in cancer prediction and prognosis. Cancer Inform. 2: 59–78.

[33] Duda, R. O., Hart, P. E. and Stork, D. G. 2001. Pattern Classification (2nd edition). New York: Wiley.

[34] William, H. W. 2020. General Surgery Dept. University of Wisconsin, Clinical Sciences Center Madison, WI 53792.

[35] Mintz, Y. and Brodie, R. 2019. Introduction to artificial intelligence in medicine. Minim. Invasive Ther. Allied Technol. 28(2): 73–81.

[36] Becker, A. 2019. Artificial intelligence in medicine: What is it doing for us today? Literature Review. HPT 8(2): 198–205.

[37] Chen, L. K. 2018. Artificial intelligence in medicine and healthcare. J. Clin. Gerontol. Geriatr. 9(3): 77–78.

[38] Scerri, M. and Grech, V. 2020. Artificial intelligence in medicine. Early Hum. Dev. 105017.

[39] Schnall, M. D., Imai, Y., Tomaszewski, J., Pollack, H. M., Lenkinski, R. E. et al. 1991. Prostate cancer: local staging with endorectal surface coil MR imaging. Radiology 178: 797–802.

[40] Ward, A. D., Crukley, C., McKenzie, C. A., Montreuil, J., Gibson, E. et al. 2012. Prostate: registration of digital histopathologic images to *in vivo* MR images acquired by using endorectal receive coil. Radiology 263: 856–864.

[41] Litjens, G., Toth, R., van de Ven, W., Hoeks, C., Kerkstra, S. et al. 2014. Evaluation of prostate segmentation algorithms for MRI: The PROMISE12 challenge. Med. Image Anal. 18: 359–373.

[42] Mangasarian, O. L., Setiono, R. and Wolberg, W. H. 1990. Pattern recognition via linear programming: Theory and application to medical diagnosis. pp. 22–31. *In*: Large-Scale Numerical Optimization; SIAM: Philadelphia, PA, USA.

[43] Wolberg, W. H. and Mangasarian, O. L. 1990. Multisurface method of pattern separation for medical diagnosis applied to breast cytology. Proc. Natl. Acad. Sci. USA 87: 9193–9196.

[44] Quinlan, J.R. 1996. Improved use of continuous attributes in C4.5. J. Artif. Intell. Res. 4: 77–90.

[45] Bennett, K. P. and Blue, J. A. 1998. A support vector machine approach to decision trees. *In*: Proceedings of the IEEE International Joint Conference on Neural Networks, Anchorage, AK, USA, 3: 2396–2401.

[46] Setiono, R. 2000. Generating concise and accurate classification rules for breast cancer diagnosis. Artif. Intell. Med. 18: 205–219.

[47] Sarkar, M. and Leong, T. Y. 2000. Application of k-nearest neighbors algorithm on breast cancer diagnosis problem. pp. 759–763. *In*: Proceedings of the AMIA Symposium, Los Angeles, CA, USA.

[48] Pach, F. P. and Abonyi, J. 2006. Association rule and decision tree based methods for fuzzy rule base generation. World Acad. Sci. Eng. Technol. 13: 45–50.

[49] Karabatak, M. and Ince, M. C. 2009. An expert system for detection of breast cancer based on association rules and neural network. Expert Sys. Appl. 36: 3465–3469.

[50] Chen, H. L., Yang, B., Liu, J. and Liu, D. Y. 2011. A support vector machine classifier with rough set-based feature selection for breast cancer diagnosis. Expert Syst. Appl. 38: 9014–9022.

[51] Koyuncu, H. and Ceylan, R. 2013. Artificial neural network based on rotation forest for biomedical pattern classification. pp. 581–585. *In*: Proceedings of the 36th International Conference on Telecommunications and Signal Processing, Rome, Italy.

[52] Kumar, U. K., Nikhil, M. B. S. and Sumangali, K. 2017. Prediction of breast cancer using voting classifier technique. *In*: Proceedings of the IEEE International Conference on Smart Technologies and Management for Computing, Communication, Controls, Energy and Materials, Chennai, India.

[53] Yue, W., Wang, Z., Chen, H., Payne, A. and Liu, X. 2018. Machine learning with in breast cancer diagnosis and prognosis. Designs 2: 13.

[54] Li, Z., Hu, Z., Xu, J., Tan, T., Chen, H. et al. 2018. Computer-aided diagnosis of lung carcinoma using deep learning—a pilot study. Computer Vision & Pattern Recognition arXiv: 1803.05471.

[55] Kourou, K., Exarchos, T. P., Exarchos, K. P., Karamouzis, M. V., Fotiadis, D. I. et al. 2015. Machine learning applications in cancer prognosis and prediction. Comput. Struct. Biotechnol. J. 13: 8–17.

[56] Chang, S. W., Abdul-Kareem, S., Merican, A. F. and Zain, R. B. 2013. Oral cancer prognosis based on clinicopathologic and genomic markers using a hybrid of feature selection and machine learning methods. BMC Bioinformatics 14: 170.

[57] Exarchos, K. P., Goletsis, Y. and Fotiadis, D. I. 2012. Multiparametric decision support system for the prediction of oral cancer reoccurrence. IEEE Trans Inf. Technol. Biomed. 16: 1127–34.

[58] Zhou, L., Wang, L., Wang, Q. and Shi, Y. (eds.). 2015. Machine Learning in Medical Imaging. Cham: Springer International Publishing. pp. 118–26.

[59] Pomponiu, V., Nejati, H. and Cheung, N.-M. 2016. Deepmole: Deep neural networks for skin mole lesion classification. pp. 2623–27. *In*: 2016 IEEE International Conference on Image Processing (ICIP). Phoenix, AZ.

[60] Kawahara, J., BenTaieb, A. and Hamarneh, G. 2016. Deep features to classify skin lesions. pp. 1397–400. *In*: 2016 IEEE 13th International Symposium on Biomedical Imaging (ISBI). Prague, Czech Republic. Available at: https://ieeexplore.ieee.org/document/7493528. Accessed 18 Dec 2019.

[61] Brinker, T. J., Hekler, A., Enk, A. H., Klode, J., Hauschild, A., Berking, C. et al. 2019. Deep learning outperformed 136 of 157 dermatologists in a head-to-head dermoscopic melanoma image classification task. Eur. J. Cancer 113: 47–544.

[62] Brinker, T. J., Hekler, A., Enk, A. H. et al. 2019. Deep neural networks are superior to dermatologists in melanoma image classification. Eur. J. Cancer. 119: 11–7.

[63] Gupta, N., Kupfer, S. S. and Davis, A. M. 2019. Colorectal cancer screening. JAMA 321: 2022–2023.

[64] Acs, B., Rantalainen, M. and Hartman, J. 2020. Artificial intelligence as the next step towards precision pathology. J. Intern. Med. 288: 62–81.

[65] Rathore, S., Hussain, M., Aksam Iftikhar, M. and Jalil, A. 2015. Novel structural descriptors for automated colon cancer detection and grading. Comput. Methods Programs Biomed. 121: 92–108.

[66] Rathore, S., Hussain, M. and Khan, A. 2015. Automated colon cancer detection using hybrid of novel geometric features and some traditional features. Comput. Biol. Med. 65: 279–296.

[67] Yang, K., Zhou, B., Yi, F., Chen, Y., Chen, Y. et al. 2019. Colorectal cancer diagnostic algorithm based on sub-patch weight color histogram in combination of improved least squares support vector machine for pathological image. J. Med. Syst. 43: 306.

[68] Mesko, B., Drobni, Z., Benyei, E., Gergely, B., Gyorffy, Z. et al. 2017. Digital health is a cultural transformation of traditional healthcare. Mhealth 3: 38.

CHAPTER 4

Applications of Swarm Intelligence and Machine Learning for COVID-19

Anurag[1] and *Naren, J*[2],*

1. Introduction

Coronavirus, also known as Severe Acute Respiratory Syndrome-Corona Virus-2 or SARS-CoV-2, is a virus strain that causes failure of the respiration function in humans. The virus is believed to have started in Wuhan, China, first detected in Dec 2019. COVID-19 virus is spread from mouth or nose droplets or direct contact with the infected person. Furthermore, recent studies show that it can be spread through air transfer (as confirmed by World Health Organization). Acute symptoms of a person infected with COVID-19 include fever, dry cough, tiredness, and some less common symptoms like sore throat, headaches, diarrhea, acne & pain, loss of taste & smell, conjunctivitis, and others [1]. The person may not show any of these symptoms in some cases but still can act as a carrier and spread the virus. Since the virus is spreading so fast and there is no cure for this disease, the most crucial task now is to detect and isolate the infected people from the healthy ones, to stop further infection. Nationwide lockdowns, quarantining, masking in public spaces, frequent hand sanitization and maintaining a social distance of 6 feet, are some of the most important steps suggested by our government to curb the corona spread [2].

---

[1] Independent Researcher, India.
[2] Senior Faculty CSE, iNurture Education Solutions Pvt Ltd, Bangalore, Karnataka, India.
Email: anurag.0496@gmail.com
* Corresponding author: naren.jeeva3@gmail.com

It is not the first time we have seen a coronavirus variant in this world, such as SARS-CoV, MERS-CoV [2]. Although different vaccines such as Pfizer, Moderna, Covaxin and Covishield are given to increase immunity against the coronavirus, and research for a cure is ongoing, the fight is far from over. The most critical step to stop this spread is detecting, diagnosing, and isolating the infected people as soon as possible. There are many reasons for the need to catch an infected person in time. The most concerning thing about this virus is that the infected person does not show symptoms immediately. It takes around 1–14 days for the body to show signs (known as the incubation period) [4]. Also, sometimes the person infected with this virus does not show any symptoms and acts like a carrier. The current method for detecting COVID, i.e., RT-PCR test [4], is long, unreliable, and takes more than 24 hours for virus detection. The infected person can cause a virus spread in this waiting period. Poor Healthcare systems worldwide, especially in developing nations like India, are not equipped with the required infrastructure to test every person and provide necessary healthcare to those infected. Finally, vaccination drives have been started in almost all countries. However, the vaccinated person can be infected with this virus and still spread it to other people.

To improve the speed and efficiency of detection, prediction and quick diagnosis of an infected person, different Machine Learning (ML) algorithms and swarm intelligence tools have been introduced [4]. Researchers introduced various ML models to classify, diagnose, detect trends, predict future curves/ forecasts, and for creating vaccines using different Artificial intelligence techniques for COVID-19. Studies have shown that the only solution to save a person's life is early detection, so that there is a possibility of complete recovery. Also, the treatment becomes difficult in later stages of detection. Machine learning is a branch of Artificial Intelligence that allows machines to learn and understand the data provided as we humans do in our day-to-day lives. The machine learning classifiers divides the raw data fed into the required categories as per the model requirement. In the past, ML has been used to classify images in the medical field and has been helpful. Swarm intelligence is the collective behavior of groups (natural or artificial) and how they react locally and to their surroundings. These groups are without a centralized control structure and simple rules which are unknown to individual agents. Swarm optimization is a tool used to optimize the previously created algorithms to increase their efficiency and performance. The SO concept is introduced with ML to improve algorithms' performances, discussed in later sections.

Before diving into the various ML and SO algorithms introduced by multiple researchers, ML classifiers and SO tools used by researchers for COVID-19 since the outbreak have been discussed in this chapter, which will be discussed in detail in later sections.

2. Data Set

For creating successful models against COVID-19, lots of data has been used by researchers to create the most optimized models to improve the detection, diagnosis, and prediction as efficiently as possible and within less time. This collected data is used to train, validate and test the models. The researchers collected the data from various sources to come up with the most optimized models. The most prominent sources of data are as follows: John Hopkins institute, USA [1], GitHub open-source data repository [21], and Kaggle. com [2]. Other researchers collected data from the hospital repository such as Albert Einstein Hospital, Sau Paulo [13], Joseph Paul Cohen [15], Sheba Medical Center [6]. For predicting and forecasting the future trends, data about number of infected cases, deaths, and recovered cases was collected from the official portals of a country, such as the Ministry of Human and Health affairs, India [5, 8], General directorate of epidemiology, secretary of Health, Mexico [10], and *Protezione Civile Italiana,* Italy [22].

2.1 Types of Data Collected

The CT scan images or the chest X-ray (CXR) images of infected persons and non-infected persons were collected for detection and classification. For prediction and forecasting the future trends, case count and time-series data of numbers infected, deaths, and recovered cases were considered for a specific number of days. The health and travel history and different factors such as age and sex, were collected to detect the severity of the infected cases. To predict the probability of recovery as per diet patterns, data on diets of people from various regions was collected.

3. Data Preprocessing

The collected data is preprocessed before sending it to train and test different ML models. The preprocessing methods help reduce noise, remove irrelevant information, convert data in one universal language for textual or time-series data, and perform gradient thresholding on CXR and CT scan datasets. Also, data augmentation processes were employed at this stage to ensure the models have sufficient data to train and test on. These preprocessing techniques preprocess data so that all the relevant data required for ML models is preserved.

4. Feature Selection/Extraction Methods

Before feeding the data to ML models, the most crucial step is to extract or select features required for prediction, forecasting and detection. In case of pictorial data, gradient thresholding is done to separate the background from an image. Then the irregularities and anomalies are detected. CNN-based architectures were used to extract the most essential features present for classification purposes [24, 25, 28, 29]. However, since it requires a large dataset and its implementation is costly, researchers used various other algorithms. Local Binary Patterns (LBP) algorithm extracts the textual features of an image and works by providing a threshold to a pixel along with each of its neighboring pixels. LBP was implemented in [12] to detect the anomaly present in CXR images. [8, 16] used the Dimensionality Reduction technique to reduce the dimensions of the feature matrix, which included features like travel history, age, sex, climate, and selected the most significant features. Label encoder preprocessing technique was used by [17] to select specific features that affect the model's overall performance. [26] proposed a hybrid selection method with Gray Level Co-occurrence Matrix (GLCM) to extract the important features for diagnosis purposes. Principal Component Analysis (PCA) is a powerful feature smoothening and reduction technique which increases the accuracy of implemented models [15, 28].

5. Machine Learning Techniques and Classifiers

Many ML models were designed and used by researchers and scientists extensively in medical science. The ML models have also been used to fight against deadly versions of the corona virus such as SARS-CoV [2], MERS-CoV [2] and ARDS [4]. The ML algorithms have been extensively used in research for diagnosis of cardiovascular disease [9], coronary heart disease [9], Diabetic retinopathy [31], stress in students [30], ASM [32], disabilities [33].

The essential ML techniques are Supervised, Unsupervised, Reinforcement and semi-supervised techniques. Researchers mostly implemented the supervised ML technique for detection, classification, forecasting and prediction of the COVID-19 pandemic. Supervised techniques work with annotated data, whereas unsupervised techniques include data that is non-annotated. The semi-supervised method works with a mixture of annotated and non-annotated data [10].

Various ML models are employed in different studies to improve the efficiency and performance of the system. These classification, forecasting, prediction and detection models were designed with the help of ML models which included individual classifiers such as Random tree (R-tree) [12], Random forest (R-forest) [3, 12], Support Vector Machine (SVM) [10, 13, 17, 26, 28], LASSO [4, 9], Decision tree [6, 10, 19], Exponential smoothing (ES) [4, 9, 16, 18], Naïve Bayes [10, 28], K-nearest neighbor (KNN) [12, 17] and many more. These models also included ensemble classifiers such as Adaptive Boosting (AdaBoost) [13], Extensive Gradient Boosting (XGBoost) [13, 20] and Bagging [3]. Ensemble models are those made when two or more machine learning models are combined while increasing the overall efficiency and performance of the model. These models perform better as compared to the individual models used to form this model. This chapter also includes some of the Deep learning algorithms such as ANN [6, 10], CNN [15, 20] and Alexnet [25], for providing better classification efficiency and performance.

Researchers and scientists have implemented various Swarm optimization models to optimize and increase the performance of ML models. Genetic Algorithm (GA) has been used in ML models previously to fight against SARS and MERS, variants of coronavirus [10]. Most of the models employed Particle swarm optimization (PSO) for feature selection [28, 29] and to optimize hyper parameters for ML models [26, 28, 29]. Guided-Grey Wolf optimizer (GWO) was used in a study for feature selection and reduction [25, 27]. Another SI algorithm, Whale optimization algorithm (WOA) [25], was used for feature selection. Marine Predators Algorithm (MPA) [23, 24] and Moth flame optimization algorithm (MFO) [23] optimized the hyper parameters for ML models to extract useful features. Improved MPA [21] algorithm found out optimized threshold values for preprocessing.

Evaluation factors are used to analyze the performance of ML models. Essential characteristics of a ML model's performance include Sensitivity/ Recall, Specificity, Accuracy, and Precision, which show the capability of the ML model to complete the specific task. A confusion matrix provides all the details about evaluation factors, including Positive Predictive Value (PPV). The prediction capacity of a model is evaluated based on the R^2 score; Root mean square Error (RMSE), Mean Absolute Error (MAE), Mean Square Error (MSE) and other measures. The optimization tools' performance is judged based on factors like F1 score, AUC (Area Under Curve), Mean Square Displacement (MSD) & Mean Absolute Percentage Error (MAPE). Many researchers have used various ML classifiers and SO tools extensively for COVID-19 since the models based on these algorithms are quickly built and require less time for training and testing. These models can perform better even with fewer data sets and have shown more than 90% accuracy.

6. Application of Machine Learning and Swarm Optimization in COVID-19

To detect, predict, forecast future trends, and diagnose the COVID-infected patients, researchers created optimized models with the help of ML algorithms and SO tools, which out performed traditional models and produced accurate and efficient results. Scientists and researchers have used various models using ML classifiers and SO tools in chronological order from May 2020 to May 2021. Punn, N. S. et al. conducted an epidemic analysis of Covid-19 to analyze the transmission growth and forecast possible transmission growth in the future [1]. They collected daily case reports and time-series summaries from Jan 22 to Apr 21, 2020, from John Hopkins University, USA. It included six attributes province/state, country/region, last update, confirmed, death, and recovered cases. Various supervised ML classifiers such as Support Vector and Polynomial regression and DNN & RNN with LSTM cells predicted the number of confirmed, death, and recovered cases for the next ten days. With least RMSE value, Polynomial regression was considered the best forecasting model. Yadav, R. S. used regression algorithms to forecast the future trend of infection cases, deaths, and recoveries and predict the transmission rate [2]. The dataset consisting of time-series data (from Jan 3 to Apr 11, 2020) from Kaggle and WHO repository was used to train various regression models like quadratic, 3rd degree, 4th degree, 5th degree, 6th degree, exponential polynomial regression models. The 6th-degree regression model forecasted cases from 12–19 April 2020 with the most negligible RMSE value and is the preferred choice for forecasting the subsequent six-day cases in India. Akib Moni et al. classified clinical test reports of 212 patients having 24 attributes such as id, age, sex, temperature and travel history, into four categories using supervised ML models [3]. These reports were collected from John Hopkins University, USA GitHub repository, and relevant features were extracted using the TF/INF technique. 70% data was used for training and 30% for testing. Various ML classifiers, such as SVM, MNB, Logistic regression, and ensemble models like bagging, Adaboost, stochastic gradient boosting forecasted the results and validated using 10-fold cross-validation strategy. LR and MNB models performed exceptionally well with 94% precision, 96% recall, 95% F1 score, and 96.2% accuracy. In the case of ensemble models, R-forest and gradient boosting achieved 94.3% accuracy. Rustam F. et al. collected daily time series summary tables from John Hopkins University GitHub repository to forecast the number of upcoming COVID patients [4]. 85% of data was used for training (56 days) and 15% for testing (10 days). Four supervised ML models Linear Regression, LASSO Regression, SVM, and ES predicted the number of newly infected cases, death rate, and recovery rate for the next ten days. ES performed very well even with the limited available data with an R2 score of 0.96 for death rate, 0.98 for newly confirmed cases, and 0.99 for future recovery rate. Also, it was observed that the performance of these models increased significantly when size of the dataset was increased. Gupta R. et al. tested SEIR and regression models to calculate the R0 value, i.e., contagiousness of the disease [5]. The data about conformed and death cases were collected from the Official portal of Government of India from Jan 30–May 10, 2020. Until May 5 (62 days), data was used to train the models, and the remaining data was used as test/evaluation data. The models predicted the no. of confirmed cases for the next 20 days and found an R0 value of 2.84, which means the disease will spread for the next 20 days, and the government needs to intervene and deploy control mechanisms. The RMSLE error rate between SEIR and regression model was 2.01 with individual RMSLE scores of 1.52 and 1.75, respectively (less is better). Assaf D. et al. predicted the patient risk for critical COVID-19 based on status at admission [6]. The data of 6995 patients from Sheba Medical Center were collected from Mar 8–Apr 5, 2020. All critical patients at the time of admission were excluded, and the 70/30 data model was followed (70% training/30% testing). Three ML algorithms ANN, R-forest, and Classification and regression decision tree (CRT), were used for prediction and validated using a 10-fold cross-validation technique. After six days of hospitalization, 29 developed severe disease, including 25 critical cases. Out of 25, 8 patients died, and others recovered. Results showed that for RFC models, APACHE II score, white blood cell count, the time from symptoms to admission, and oxygen saturation were contributory factors. In contrast, for the ANN model, blood lymphocytes contributed more than oxygen saturation. The ML models increased the accuracy (more than 90%) and discriminative efficiency of these factors. Andreas A. et al. used the time-series data of 105 days from ourworldindata.org starting from Jan 22, 2020 [7]. Three ML models, linear, polynomial, and 6th-degree regression models, were tested on this data. It was found that the 6th-degree regression model performed more accurately with an R2 score of 0.99939. Moreover, to predict the trend and growth of COVID, cloud computing was introduced with ML models. Levensburg Marquardt algorithm was used for producing the curve fitting model for Italy. 10th-degree polynomial was developed in MATLAB for curve fitting. These forecast models helped predict live cases count and aided the government to avoid the situation of premature uplifting of lockdown and formulate their policies as early as possible. Gambhir E. et al. performed a regression analysis of time series data of confirmed, death, and recovered cases of 154 days collected from mohfw.gov.in, the official portal of India and John Hopkins University from Jan 22–Jun 24, 2020 [8]. Important features were extracted using the DR feature extraction technique, and supervised ML algorithms SVM and polynomial regression (PR) were tested. PR attained an accuracy of 93% predicting values from Aug 7–Aug 28, 2020. They also conducted a case study in India state/UTs wise to realize the current and future trends using the above models and forecasted the increase in cases for the next 60 days. These trends will help the government of hugely populated countries like India to take action early. Kumar R. et al. designed supervised ML models for future forecasting [9]. The dataset of confirmed cases, deaths, and recoveries was collected from John Hopkins University, USA GitHub repository from Jan 2020 till Oct 2020. 85% data was used for training and 15% data for testing. Four supervised ML models, Liner, regression, LASSO, SVM, and ES were tested and the prediction forecast for the next ten days was calculated. ES performed best for predicting number of recoveries and newly confirmed cases, whereas LASSO and LR predicted best results for number of deaths for the next ten days.

L.J. Mohammad et al. predicted the COVID-19 infection using the data collected from the General Directorate of Epidemiology, Secretariat of Health, Mexico [10]. This dataset included 263,007 records with 41 features (only two demographic features: age and sex and eight clinical features like pneumonia, diabetes, asthma, obesity along with one high-risk factor - tobacco were extracted), and 80% of data was used to train the models and 20% data to test. Various supervised ML algorithms such as Naïve Bayes, Logistic Regression, Decision tree, SVM, and ANN were tested on preprocessed data and were evaluated based on accuracy, specificity, and sensitivity. Decision tree attained highest - 94.99% accuracy, SVM highest - 99.34% sensitivity, and Naïve Bayes highest - 94.3% specificity. Furthermore, while accessing the features using Decision Tree, it was found that age was the most significant feature for COVID infection, and males were more prone to get infected than females. Hossen, M. S. and Karmoker D. predicted the recovery probability in various countries based on healthy diet patterns [11]. Since the recovery from COVID-19 depends upon the immunity power, which depends upon food habits. The dataset was collected from Kaggle repository and worldometer website with diet patterns from South Asian countries. Some of the essential features like country, Alcohol beverages, animal products, cereals, recovered, and the death count were extracted. These various food consumptions of South Asian countries were compared with the top 10 COVID-affected countries. The ML models R-forest, SVM, & KNN predicted that more plant-based products and less animal-based products in diet patterns significantly increased the recovery rate. Thepade, S. D. et al. used local binary patterns for anomaly detection in chest X-ray images collected from the Kaggle repository [12]. The data included 68 Corona infected CXRs, 79 Normal CXRs, and 158 pneumonia affected CXRs. This data was classified using supervised ML models like Naïve Bayes and KNN, and ensemble models like R-tree-R-forest-KNN and R-forest-R-tree-SVM validated with a 10-fold cross-validation technique. R-tree-R-forest-KNN attained the highest accuracy of 89.18%, average sensitivity of 0.892, and average F-measure of 0.888. Furthermore, the results showed that the ensemble models performed better than individual classifiers. Hamide S. et al. optimized the hyper parameters for ML models to improve their performance [13]. The dataset containing the viral and Blood data of 1260 patients was collected from Albert Einstein Hospital, Sau Paulo, Brazil. The ratio used for the training and testing process was 80% and 20%, respectively. The evaluation parameters of five ML models, namely SVM, AdaBoost, R-forest, XGBoost, and Decision Tree, were compared using the Grid search function. The study concluded that optimized parameter AdaBoost algorithm's recall increased by 18%, followed by SVM (14% increase in recall). Therefore, Grid search function can be used to improve the performance of already existing models. Dharani N. P. et al. evaluated the performance of ML models Linear Regression and SVM to predict confirmed, active, migrated, and death cases [14]. The time-series data was collected from Kaggle from Jan 30–May 21, 2020. The supervised ML models predicted the cases from 21st May–25th June 2020. The evaluation table was formulated as per RMSE, R2 score, training, and prediction speed, and the study concluded that Linear Regression model out performed the SVM model in all the prediction scenarios. These models helped to predict the amount of infection expansion for the next 25 days. Rasheed J. et al. designed a diagnosis framework based on ML models for COVID-19 detection [15]. The dataset consists of 198 infected CXRs and 210 normal CXRs collected from Joseph Paul Cohen hospital. Data augmentation was done using Generative Adversarial Analysis (GAN) for a sample size of 500 (250 infected + 250 normal), and it was converted into a 75% training set and 25% testing set. The features were selected based on PCA, and the accuracy of ML models logistic regression and CNN were compared with and without PCA. This logistic regression accuracy increased from 95.2% to 97.6%, and the CNN model attained 100% accuracy with PCA. Gera S. et al. performed regression analysis on 292 days (from 30 Jan–16 Nov 2020) of time series data collected from John Hopkins University, USA, to predict the number of newly infected, deaths, and recovered cases [16]. The dataset was converted into an 80% training set (233 days) and a 20% testing set (59 days). Important Features were selected using the dimensionality reduction technique and supervised ML models such as linear regression, SVM, KNN, R-forest, and ES. Based on an R2 score of 0.976, ES performed best among all algorithms, followed closely by R-forest (R2 score of 0.966). The forecasted results showed an increase in all cases, including the number of deaths. Also, the performance of these ML models increased with an increase in size of the dataset. Rohini M. et al. predicted using data of 11435 patient records from the Kaggle repository with 20 symptoms, which were selected using the feature extraction label encoder preprocessing technique [17]. This dataset separated as 80% training and 20% testing was supplied to supervised ML models, including KNN, SVM, Decision tree, and R-forest, which resulted in a confusion matrix for each model. The proposed KNN model performed best with 98.34% accuracy, 97% recall, and F1 score of 0.97, and the SVM model attained the best precision (97%). Gothai, E. et al. used Holt winter's ES model to predict the trend and growth of COVID [18]. The 172,479 documents from John Hopkins University, USA, included eight attributes province, country, last update, confirmed, death, recovered cases and others between Jan–Dec, 2020. Feature extraction was performed to extract recovered, confirmed, and death cases for classification. Supervised ML models linear regression, SVM, and Holt-

Winter time series ES model were trained and tested. Holt's ES outperformed all other models and achieved an accuracy of 87%. Malki Z. et al. estimated the spread of infection in many countries using ML models and the expected period after which spread of the virus will be stopped [19]. Time-series data of the top 12 countries, including USA, India, Brazil among others were collected from John Hopkins University, WHO, and the worldometer website. Supervised ML Decision tree model forecasted the values of confirmed, death, and recovery cases for the next seven days in the USA. This out performed state-of-the-art ML models based on R-forest, ARIMA, and DL models. They also estimated that the COVID-19 infection would decline in the first week of Sep 2021 and stop soon after. This model was compared with previous pandemics and proved its effectiveness. This model can further be used to forecast case count in other countries of the world. Habib N. and Rahman M. M. solved diagnosis of COVID from a different approach [20]. They introduced a gene-based screening method to find functional semantic similarities among genes and based on the results, ML models will classify these as COVID positive or not. The dataset for this research was collected from NCBI gene database and was used to train and test various ML models such as XGBoost, Naïve Bayes, Regularized R-forest and MLP. The ensemble model was designed from best performing models detected from disease genes or CT or CXR images with 93% accuracy. They proposed another model based on the deep CNN technique, which detected COVID from the dataset collected from the COVID-19 Radiography Dataset by Tawisfur Rahman (available on Kaggle). Using the deep CNN technique, the dataset consisting of COVID and healthy CXR images detected COVID-19 with 99.87% accuracy, and another dataset consisting of COVID and Pneumonia CXR detected COVID-19 with 99.48% accuracy.

6.1 Introduction to SO Tools with ML Models

Abdel Basset M. et al. proposed a model with an improved Marine predator algorithm (IMPA) with ranking-based diversity reduction (RDR) technique [21]. The dataset consisted of 9 CXR images from the GitHub repository, and IMPA was applied for data segmentation. The RDR technique was used to improve the performance of IMPA to find the optimal threshold value for the multilevel thresholding process. With a low peak-to-signal noise ratio (PSNR) and high Structured Similarity Index Matrix (SSIM) [21], the IMPA-RDR model outperformed all other previously chosen algorithms for data segmentation, giving better consistency and stability for high threshold values. Godio, A. et al. introduced a stochastic approach using PSO along with SEIR model for investigating the epidemic in Italy [22]. The dataset consists of the number of infected recovered and deaths collected from John

Table 1. Machine Learning models for prediction, forecasting, classification and detection of COVID-19.

Ref.	Type	Data Set & source	ML classifier &/or SO tools	Results & Discussion
[1]	Forecasting	John Hopkins University, USA Daily case reports and daily time series summary tables. From Jan–April 20	Support Vector Regression, Polynomial Regression, DNN, Recurrent Neural Network (RNN)	PR approach produces the best forecast results
[2]	Forecasting	Kaggle.com, WHO website Mar–April 2020 Cases Data	Quadratic, 3rd degree, 4th degree, 5th degree, 6th degree, Exponential polynomial regression	A sixth-degree regression model with least RMSE produced better forecasting results
[3]	Forecasting, Classification	John Hopkins University, USA, Open Source data repository GitHub, 212 Patient clinical test reports	SVM, MNB, Logistic Regression, MNB, Decision tree, Random forest, bagging, AdaBoost, stochastic gradient Boosting	LR & MNB Precision-94% Recall-96%, F1 score- 95%, Accuracy-96.2% R-forest, gradient boosting: Accuracy-94.3%
[4]	Forecasting	John Hopkins University, USA, GitHub repository	LR, LASSO, SVM, ES	ES performs best with an R2 score of 0.96, 0.98 & 0.99 for death rate, newly confirmed cases, and recovery rate, respectively.
[5]	Prediction	Ministry of health and family welfare, India & John Hopkins University, USA Cases data from Jan–May 2020	SEIR model, regression model	R0 value of 2.84
[6]	Detection	Sheba Medical Center (6 days) 6695 patients data	Random Forest, Classification and regression decision tree (CRT), Artificial Neural Network	ML models increased the overall accuracy of the proposed prediction modal.
[7]	Forecasting	ourworldindata.org Open access data repository	Linear, exponential, polynomial (6th degree) regression models	A 6th-degree regression model has the best R2 value (0.99936), thus preferred for forecasting
[7]	Prediction	Same as above	Levenberg-Marquardt algorithm	Live confirm cases count were predicted
[8]	Prediction	Ministry of Health and Family Welfare, India, John Hopkins University, USA	SVM, Polynomial regression (PR)	PR shows 93% accuracy and predicted increased cases in the next 60 days
[9]	Forecasting	John Hopkin's University, USA	ES, LASSO, SVM, LR	ES performs best for forecasting all cases.

Table 1 contd. ...

...Table 1 contd.

Ref.	Type	Data Set & source	ML classifier &/or SO tools	Results & Discussion
[10]	Prediction	General Directorate of epidemiology, Secretariat of Health, Mexico 263,007 records (41 features)	LR, Decision tree, SVM, Naïve Bayes, ANN	DT - 97% accuracy SVM - 99.34% recall Naïve Bayes - 94.34% specificity. Age is the biggest factor.
[11]	Prediction	COVID-19 Healthy diet dataset Kaggle. com	Random Forest SVM, KNN	More plant-based products and less animal-based products result in more recovery rates.
[12]	Classification	CXR images (68 Corona infected 79 normal lungs 158 pneumonia Patients)	Naïve Bayes, KNN, SVM, R-tree, R-forest, Ensemble models: R-forest-R-tree-SVM, R-forest-R-tree-KNN, R-forest-SVM-KNN	R-tree-R-forest-KNN performed best
[13]	Prediction	Albert Einstein Hospital, Sau Paulo	SVM, AdaBoost, Random Forest, XGBoost, Decision tree	Recall has been improved by 18% (AdaBoost) and 14% (SVM)
[14]	Forecast	Kaggle.com Jan-May 2020	LR, SVM	LR is the preferred choice (as its RMSE is less)
[15]	Detection	Joseph Paul Cohen (198 infected + 210 non-infected CXR images	LR, CNN	Accuracy: Without PCA LR-95.2%, CNN-97.6% With PCA LR-97.6%, CNN-100%
[16]	Prediction	Kaggle, GitHub repository of John Hopkins University (292 days data) Time	LR, SVM, Random forest, KNN, ES	ES performs best in predicting no. of newly confirmed cases, no. of death, and recoveries.
[17]	Prediction	Kaggle 11435 records (20 symptoms)	KNN, SVM, Decision tree, Random forest	KNN performed best with 98.34% accuracy, 97% recall, 0.97 F1 score SVM perform best w.r.t. precision (97%)

[18]	Prediction	John Hopkin's University repository (Jan–Dec 2020) 172,479 datasets	LR, SVM, ES, Holt winter ES	Holt winter ES performed best with 87% accuracy
[19]	Prediction	John Hopkins University, USA Worldometer.com Top 12 countries COVID curves (Jan 2020–Jan 2021)	Decision tree,	Outperformed previous models, predicted the end of infection in Sep 2021 in the USA.
[20]	Detection	NCBI gene database, COVID-19 Radiography Dataset by Tawsifur Rahman (Kaggle)	XGBoost, Naïve Bayes, Regularized R-forest, R-forest rule-based model, R-ferns, C5.0, MLP, deep CNN	Ensemble models: accuracy-93% CNN technique: COVID vs. normal: 99.87% accuracy COVID vs. Pneumonia: 99.48% accuracy

Hopkins University, USA. The Italian data was collected from *Protezione Civile Italiana* and population count from ISTAT. The PSO algorithm was used to fit the model parameters of the proposed model. The model was tested for predicting cases for the next 30 days and predicted the maximum number of quarantined people in various regions of Italy. The model was evaluated using the NRMSE value and compared it with the deterministic approach used before this proposed model. The NRMSE value of 0.035 was obtained for Italy, and the same model was applied to Spain and the South Korean epidemic, which resulted in NRMSE values of 0.046 & 0.074, respectively. The study concluded that the stochastic approach using the PSO algorithm showed better results than previous models.

Elaziz, M. A. et al. proposed a combination of Marine Predators Algorithm (MPA) with the Moth flame optimization (MFO) algorithm called MPAMFO algorithm to help the Multilevel threshold process of image segmentation find the optimal threshold value [23]. The study included two types of datasets; the first set consisted of 10 greyscale images for experiment 1 and 13 CT images of COVID-19 collected from various sources such as Chexpert, Opert, google, PadChest and others. The results of MPAMFO algorithms outperformed all other optimization algorithms like GWO, PSO, MPA, MFO among others, in terms of PSNR value, SSIM, and fitness number for various threshold values. The robustness of the proposed model using the Friedman test showed its superior performance over other models. The MPAMFO performed well and was stable for both experiments. Sahlot, A. T. et al. designed classification model based on advanced CNN architecture Inception [24]. They selected two datasets for their research; dataset 1 consisted of 200 positive and 1675 negative COVID CXR images collected from International Cardiologist radiologists, researchers, and images from Kaggle. Dataset 2 consisted of 219 positive and 1341 negative COVID CXR images collected from researchers and the Italian Society of Medical & Interventional Radiology (SIRM) database. The CNN-based Inception model was used for feature extraction to extract 130 features for Dataset 1 and 86 features for Dataset 2, out of 51k features. The significant features were collected using the Fractal order calculus-Marine predators' algorithm. This hybrid classification model attained 98.7% accuracy and 98.2% F score for Dataset 1 and 99.6% accuracy, and 99% F score for Dataset 2. This model successfully selected small features with high accuracy and outperformed various other similar models.

El-Kenawy et al. proposed a model with novel features selection and voting classifier algorithms to classify COVID [25]. The dataset included 334 CT COVID and 704 CT non-COVID images collected from medRziv, bioRxiv, NEJM, JAMA, & Lamcet. They selected CT scans to detect the severity of COVID as classification-based CT scan images are better than chest X-ray images [25]. The dataset was decided as 60% training set, 20% validation set, and 20% testing set. Data balancing is done using the data augmentation technique LSH-SMOTE, and specific features were extracted using CNN-based algorithm Alexnet. The guided Whale optimization algorithm (WOA) based on stochastic fractal search (SFS) has been used for features selection. Guided WOA-SFS outperformed other models like WOA, GWO, GA, PSO, GWO-PSO, and others, with a minimum average error of 0.1381. The data was processed using an ensemble model based on the best-performing SVM, NN, KNN, and Decision Tree groups. To classify the data into various classes, PSO was introduced with guided WOA as a voting classifier. This voting classifier, compared to other classifiers such as WOA, GWO, GA, PSO, among others, achieved a better result as per statistical tests like Wilconn rank-sum [25], ANOVA [25], and T-test [25] and an AUC score of 0.995. The proposed Alexnet model outperformed other models with a F-score of 0.7788.

Mohammed S. N. et al. performed a diagnosis for COVID-19 with the help of PSO [26]. The dataset of 51 CT scan chest images collected from Kaggle was used for performing computer-aided diagnosis. The essential features were extracted using Stack Hybrid Classification System & GLCM feature extraction algorithm, and automatic data segmentation was achieved using Li's method and PSO, which were used to determine the optimal threshold values. These processed images were classified using the ensemble ML model based on SVM, resulting in 98.4% accuracy, 100% sensitivity, and 91.67% specificity. The study also concluded that the accuracy of ML model decreases by 4% without PSO. Arbabili, S. F. et al. used alternate algorithms than SEIR/SIR model to predict the outbreak of COVID [27]. The dataset collected included the time series data of 5 countries for 30 days from worldometer website. The ML models such as Multilayered Perception (MLP), Adaptive Network-based fuzzy interface system (ANFIS) based on Artificial Neural Network model were introduced for future study only. In this study, researchers have introduced various optimization tools such as GA, PSO, and Grey Wolf optimizer (GWO) to optimize the hyper parameters for ML models. After evaluating different optimized ML models based on R2 value, RMSE, Correlation co-eff., among others, GWO was found to be the best optimizing tool to increase the accuracy of ML models. The logistic regression model optimized with GWO forecasted the best results for selected countries. The study concluded that the best way to forecast future trends was to integrate SEIR and ML models.

Asghar M. A. et al. implemented the 2D-CNN technique for feature extraction and PSO algorithm for feature selection to get better results than other proposed models [28]. The data was collected as two datasets from Kaggle; Dataset 1 included 6000 radiograph images of healthy, pneumonia infected, among others, whereas Dataset 2 had 900 radiographs of images of healthy and SARS, pneumonia, COVID affected images. The feature extraction was completed using a 2D-CNN algorithm, followed by feature selection using PSO. The PCA technique was implemented for feature smoothening and reduction. Before sending the data for classification using ML models such as SVM, K-NN, Naïve Bayes, and ensemble, the data was converted as non-separable time-frequency representation images using nSWT [28]. The confusion matrices of various ML models were used to find the accuracy recall, precision, specificity, and F1 score. The SVM ensemble model resulted in 99.81% accuracy better than any other available model. Hamdy W. et al. designed a model using CNN and PSO algorithms [29]. The dataset for the proposed model was collected as 16,756 CT chest images across 13,645 patients collected from 2 sources. The dataset was divided into 80% training set (20% used as validation group) and 20% testing set. A sample size of 1500 healthy and 1050 COVID infected images was selected to feed into the CNN model. The Optimal hyperparameter for CNN was selected using the PSO algorithm, and the confusion matrix was obtained. It can be seen that the accuracy of the CNN model increases from 97.17% to 98.04% with PSO. The study concluded that transfer learning based models could be designed based on their study in the future.

The overview of all the SO models designed by various researchers and scientists to fight against this global pandemic is shown in Table 2.

7. Discussion

This chapter discusses various machine learning models designed to detect the early presence of COVID-19 in patients and predict future trends accurately and efficiently. These algorithms were preferred due to their efficiency and cost-effective nature. We also covered how swarm intelligence optimization algorithms increased the overall performance of these ML models by selecting optimal or hyperparameters. These SO tools also helped in data preprocessing, feature extraction, and feature selection processes. These Models can be used as a reference by government, hospitals, and policy makers to make resource and supply management decisions, opening lockdown restrictions, vaccine supply management, and proceedings. These methods will help save lives and facilitate effective resource management.

8. Conclusion

Many developed countries like South Korea, America, Russia, etc., have flattened the curve of COVID-19 related causalities by intelligently providing vaccines to their population. Surprisingly, China, where the outbreak started, has eliminated the effect of this deadly virus with effective planning, lockdown restrictions, public surveillance techniques, etc., to ensure the safety of its large population. Whereas in a developing country such as India, the effect

Table 2. Intermingling Swarm Optimization and Machine Learning algorithms for detection, classification and forecasting of COVID-19.

Ref.	Type	Data set and source	ML classifier and/or SO tools	Results and discussions
[21]	Detection	9 CXR images from the GitHub repository	IMPA + RDR	Less PSNR value and high SSIM
[22]	Diagnosis	John Hopkins University, USA *Protezione Civile Italiana and ISTAT*	SEIR model + PSO	NRMSE = 0.035 for Italy Better performance than previous models
[23]	Detection	Chexpert, Opert, google, PadChest 13 CT images	MPAMFO algorithm	Showed superior performance
[24]	Classification	D1: 200 positive + 1675 negative COVID images from Kaggle D2: 219 positive + 1341 negative COVID images from SIRM	FO-MPA with CNN model	D1: 98.7% accuracy, 98.2% F score D2: 99.6% accuracy, 99% F score
[25]	Classification	medRziv, bioRxiv, NEJM, JAMA, & Lamcet 334 positive + 704 negative CT images	Alexnet CNN architecture with guided WOA-SFS and WOA-PSO	Alexnet F score = 0.7788 With guided WOA-SFS and guided WOA-PSO
[26]	Detection	Kaggle repository 51 CT chest images (39 COVID + 12 normal)	Ensemble SVM using Li's method and PSO	98.04% accuracy 100% sensitivity 91.6% specificity
[27]	Forecast	Worldometer.org Time series data from 5 countries	Logistic regression, polynomial regression with GWO	A logistic regression model with GWO predicted the best outcome for selected countries
[28]	Detection	Kaggle, D1:6000 radiographs (normal, pneumonia) D2:900 radiographs (COVID, pneumonia, normal, SARS)	2D CNN + PSO, SVM, K-NN, Naïve Bayes and ensemble	SVM ensemble performed best with accuracy = 99.81%
[29]	Classification, detection	16,756 CT images across 13,645 patients (1500 normal and 1050 COVID images)	CNN + PSO	Accuracy of CNN model increases from 97.17% to 98.04% with PSO

of the second wave has shaken the very foundations of countries GDP and poor healthcare systems. The vaccination drive is still in progress, and the fight against Corona is far from over. The primary purpose of this chapter is to guide future researchers to develop even better models by keeping these models as references.

References

[1] Punn, N. S., Sonbhadra, S. K. and Agarwal, S. 2020. COVID-19 epidemic analysis using machine learning and deep learning algorithms. MedRxiv. https://doi.org/10.1101/2020. 04.08.20057679.

[2] Yadav, R. S. 2020. Data analysis of COVID-2019 epidemic using machine learning methods: a case study of India. International Journal of Information Technology (Singapore) 12(4): 1321–1330. https://doi.org/10.1007/s41870-020-00484-y.

[3] Khanday, A. M. U. D., Rabani, S. T., Khan, Q. R., Rouf, N. and Mohi Ud Din, M et al. 2020. Machine learning based approaches for detecting COVID-19 using clinical text data. International Journal of Information Technology (Singapore) 12(3): 731–739. https://doi.org/10.1007/s41870-020-00495-9.

[4] Rustam, F., Reshi, A. A., Mehmood, A., Ullah, S., On, B. W. et al. 2020. COVID-19 future forecasting using supervised machine learning models. IEEE Access 8: 101489–101499. https://doi.org/10.1109/ACCESS.2020.2997311.

[5] Gupta, R., Pandey, G., Chaudhary, P. and Pal, S. K. 2020. Machine learning models for government to predict COVID-19 outbreak. Digital Government: Research and Practice 1(4): 1–6. https://doi.org/10.1145/3411761.

[6] Assaf, D., Gutman, Y., Neuman, Y., Segal, G., Amit, S. et al. 2020. Utilization of machine-learning models to accurately predict the risk for critical COVID-19. Internal and Emergency Medicine 15(8): 1435–1443. https://doi.org/10.1007/s11739-020-02475-0.

[7] Andreas, A., Mavromoustakis, C. X., Mastorakis, G., Mumtaz, S., Batalla, J. M. et al. 2020. Modified machine learning technique for curve fitting on regression models for COVID-19 projections. IEEE International Workshop on Computer Aided Modeling and Design of Communication Links and Networks, CAMAD, 2020-September (December 2019). https://doi.org/10.1109/CAMAD50429.2020.9209264.

[8] Gambhir, E., Ritika, J., Gupta, A. and Tomar, U. 2020. Regression analysis of COVID-19 using machine learning algorithms. Proceedings of the International Conference on Smart Electronics and Communication (ICOSEC 2020).

[9] Kumar, R., Yadav, A., Prabhu, A. V. and Natarajan, Y. 2020. Since January 2020 Elsevier has created a COVID-19 resource centre with free information in English and Mandarin on the novel coronavirus COVID-19. The COVID-19 resource centre is hosted on Elsevier Connect, the company' s public news and information website. Elsevier hereby grants permission to make all its COVID-19-related research that is available on the COVID-19 resource centre—including this research content—immediately available in PubMed Central and other publicly funded repositories, such as the WHO COVID database with rights for unrestricted research re-use and analyses in any form or by any means with acknowledgement of the original source. These permissions are granted for free by Elsevier for as long as the COVID-19 resource centre remains active. Machine learning models for covid-19 future forecasting (January).

[10] Muhammad, L. J., Algehyne, E. A., Usman, S. S., Ahmad, A., Chakraborty, C. et al. 2021. Supervised machine learning models for prediction of COVID-19 infection using

epidemiology dataset. SN Computer Science 2(1). https://doi.org/10.1007/s42979-020-00394-7.

[11] Hossen, M. S. and Karmoker, D. 2020. Predicting the probability of Covid-19 recovered in South Asian countries based on healthy diet pattern using a machine learning approach. 2020 2nd International Conference on Sustainable Technologies for Industry 4.0, STI 2020, 0: 19–20. https://doi.org/10.1109/STI50764.2020.9350439.

[12] Thepade, S. D. and Jadhav, K. 2020. Covid19 identification from chest x-ray images using local binary patterns with assorted machine learning classifiers. 2020 IEEE Bombay Section Signature Conference, IBSSC 2020: 46–51. https://doi.org/10.1109/IBSSC51096.2020.9332158.

[13] Hamida, S., Gannour, O. E. L., Cherradi, B., Ouajji, H., Raihani, A. et al. 2020. Optimization of machine learning algorithms hyper-parameters for improving the prediction of patients infected with COVID-19. 2020 IEEE 2nd International Conference on Electronics, Control, Optimization and Computer Science, ICECOCS 2020 (1). https://doi.org/10.1109/ICECOCS50124.2020.9314373.

[14] Dharani, N. P., Bojja, P. and Raja Kumari, P. 2021. Evaluation of performance of an LR and SVR models to predict COVID-19 pandemic. Materials Today: Proceedings (xxxx). https://doi.org/10.1016/j.matpr.2021.02.166.

[15] Rasheed, J., Hameed, A. A., Djeddi, C., Jamil, A., Al-Turjman, F. et al. 2021. A machine learning-based framework for diagnosis of COVID-19 from chest X-ray images. Interdisciplinary Sciences: Computational Life Sciences 13(1): 103–117. https://doi.org/10.1007/s12539-020-00403-6.

[16] Malki, Z., Ewis, A., Atlam, E., Degnew, G., Dqg, H. et al. 2021. The COVID-19 pandemic: prediction study based on machine learning models. Environmental Science and Pollution Research, https://doi.org/10.1007/s11356-021-13824-7.

[17] Rohini, M., Naveena, K. R., Jothipriya, G., Kameshwaran, S., Jagadeeswari, M. et al. 2021. A comparative approach to predict corona virus using machine learning. Proceedings—International Conference on Artificial Intelligence and Smart Systems, ICAIS 2021, (Ml), 331–337. https://doi.org/10.1109/ICAIS50930.2021.9395827.

[18] Gothai, E., Thamilselvan, R., Rajalaxmi, R. R., Sadana, R. M., Ragavi, A. et al. 2021. Prediction of COVID-19 growth and trend using machine learning approach. Materials Today: Proceedings, (xxxx). https://doi.org/10.1016/j.matpr.2021.04.051.

[19] Malki, Z., Atlam, E. S., Ewis, A., Dagnew, G., Ghoneim, O. A. et al. 2021. The COVID-19 pandemic: prediction study based on machine learning models. Environmental Science and Pollution Research. https://doi.org/10.1007/s11356-021-13824-7.

[20] Habib, N. and Rahman, M. M. 2021. Diagnosis of corona diseases from associated genes and X-ray images using machine learning algorithms and deep CNN. Informatics in Medicine Unlocked 24: 100621. https://doi.org/10.1016/j.imu.2021.100621.

[21] Abdel-Basset, M., Mohamed, R., Elhoseny, M., Chakrabortty, R. K., Ryan, M. et al. 2020. A hybrid COVID-19 detection model using an improved marine predators algorithm and a ranking-based diversity reduction strategy. IEEE Access 8: 79521–79540. https://doi.org/10.1109/ACCESS.2020.2990893.

[22] Godio, A., Pace, F. and Vergnano, A. 2020. Seir modeling of the italian epidemic of sars-cov-2 using computational swarm intelligence. International Journal of Environmental Research and Public Health 17(10). https://doi.org/10.3390/ijerph17103535.

[23] Elaziz, M. A., Ewees, A. A., Yousri, D., Alwerfali, H. S. N., Awad, Q. A. et al. 2020. An improved marine predators algorithm with fuzzy entropy for multilevel thresholding: real world example of COVID-19 CT image segmentation. IEEE Access 8: 125306–125330. https://doi.org/10.1109/ACCESS.2020.3007928.

[24] Sahlol, A. T., Yousri, D., Ewees, A. A., Al-qaness, M. A. A., Damasevicius, R. et al. 2020. COVID-19 image classification using deep features and fractional-order marine predators algorithm. Scientific Reports 10(1): 1–15. https://doi.org/10.1038/s41598-020-71294-2.

[25] El-Kenawy, E. S. M., Ibrahim, A., Mirjalili, S., Eid, M. M., Hussein, S. E. et al. 2020. Novel feature selection and voting classifier algorithms for COVID-19 classification in CT images. IEEE Access 8. https://doi.org/10.1109/ACCESS.2020.3028012.

[26] Mohammed, S. N., Alkinani, F. S. and Hassan, Y. A. 2020. Automatic computer aided diagnostic for COVID-19 based on chest X-Ray image and particle swarm intelligence. International Journal of Intelligent Engineering and Systems 13(5): 63–73. https://doi.org/10.22266/ijies2020.1031.07.

[27] Ardabili, S. F., Mosavi, A., Ghamisi, P., Ferdinand, F., Varkonyi-Koczy et al. 2020. COVID-19 outbreak prediction with machine learning. Algorithms 13(10). https://doi.org/10.3390/a13100249.

[28] Asghar, M. A., Razzaq, S., Rasheed, S. and Fawad. 2020. A robust technique for detecting SARS-CoV-2 from X-ray image using 2D convolutional neural network and particle swarm optimization. 2020 14th International Conference on Open Source Systems and Technologies, ICOSST 2020—Proceedings. https://doi.org/10.1109/ICOSST51357.2020.9333084.

[29] Hamdy, W., Elansary, I., Darwish, A. and Hassanien, A. E. 2021. An optimized classification model for COVID-19 pandemic based on convolutional neural networks and particle swarm optimization algorithm. Studies in Systems, Decision and Control 322(March): 43–61. https://doi.org/10.1007/978-3-030-63307-3_3.

[30] Naren, J. and Vithya, G. 2019. Predicting academic performance in students by an analytical study on big data machine learning techniques. International Journal of Psychosocial Rehabilitation 23(1): 371–376. https://doi.org/10.37200/ijpr/v23i1/pr190247.

[31] Hemalatha, R., Anjanadevi, V., Naren, J. and Vithya, G. 2019. A detailed study on diagnosis and prediction of diabetic retinopathy using current machine learning and deep learning techniques. International Journal of Psychosocial Rehabilitation 23(1): 412–417. https://doi.org/10.37200/ijpr/v23i1/pr190253.

[32] Sairam, K., Naren, J., Vithya, G. and Srivathsan, S. 2019. Computer aided system for autism spectrum disorder using deep learning methods. International Journal of Psychosocial Rehabilitation 23(1): 418–425. https://doi.org/10.37200/ijpr/v23i1/pr190254.

[33] Rakshanasri, S. L., Naren, J., Vithya, Dr G., Akhi, S., Dinesh Kumar, K. and Sai Krishna, S. 2020. Health smart home with IoT—A state of art survey. International Journal of Psychosocial Rehabilitation 24(04): 165–175. https://doi.org/10.37200/ijpr/v24i4/pr200996.

CHAPTER 5

Machine Learning for Rural Healthcare

Parveen Kumar Lehana,[1] *Chaahat,*[2,]* *Akshita Abrol,*[1] *Priti Rajput*[1] and *Parul*[1]

1. Introduction

The global pandemic of the novel Corona virus in the 21st century, human health across countries worldwide has been adversely affected. Machine Learning (ML) is taking control over the traditional healthcare system by providing automated, faster and efficient analysis of medical problems. There is enough confirmation in support of the positive transformations, machine learning has made in the healthcare industry and human health. Chatbots is the finest example of machine learning for healthcare application where a virtual nurse in the form of a voice controlled healthcare assistant provides information on patient's medical data.

There are numerous other areas in healthcare where machine learning finds application. It is used in disease diagnosis, such as genetic and cancer diseases, an exemplification of which is IBM Watson Genomics. I-Assistant is another example of a machine learning solution in healthcare that provides visually impaired and blind people an alternative to an in-person test writer and reader in an exam, giving them an equal opportunity to pursue their careers. Various other applications of machine learning in healthcare include fractal-based speech analysis for estimation of human emotional content, diabetes and liver disease prediction.

[1] DSP Lab, Department of Electronics, University of Jammu, J&K, India.
[2] Assistant Professor, Model Institute of Engineering and Technology, Jammu, J&K, India.
Email: pklehana@gmail.com
* Corresponding author: chaahatgupta249@gmail.com

Shortage of healthcare workers in rural areas has led to the advancement of artificial intelligence (AI) based tele-health systems where the patients can get real time health monitoring via remote interactions with experts in their respective fields. Thus, automation using machine learning algorithms benefits the healthcare system by saving costs incurred on doctors visit and patient hospitalizations. AI and ML have found beneficial application for healthcare purposes such as robotic surgery, electroceuticals, personalized medicine recommendations and Covid disease analysis.

Machine learning in healthcare has tremendously hastened the process of sorting and classifying health data, thus, helping doctors to make timely predictions on diseases, thus saving many valuable lives. ML in healthcare provides various technologies that contribute to the future of advanced medical diagnostics. It is feared that the advancement of machine learning in healthcare will replace human clinicians in the near future, but rather will supplement and augment their efforts to take better care of patients. Today, there are many health-related applications utilizing the combined power of machine learning with that of a remote physician for timely treatment of patients. The future of ML in the healthcare industry is very promising and welcome.

Research on the development of smart bandaging is being carried at the Digital signal processing lab in the Department of Electronics, Jammu University. Machine based techniques are being used to develop smart bandages at ease in a short time duration. By capturing the image of the wound, the CNN based smart system develops the design of the smart bandage system and the flexible bandage can be printed within seconds using Arc Sign Printer.

Another group of researchers at the Digital signal processing lab in the department of Electronics, Jammu University is working on developing techniques for knowing the emotional state of the speaker. The same technology can be used in cell phones so that a person sitting at a distant place

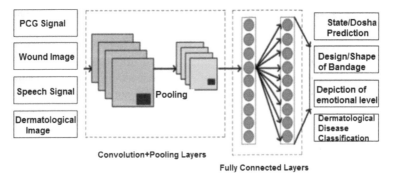

Fig. 1. Applications of convolutional neural network in healthcare.

can be emotionally helped and it may also warn the listener of the stress level of the speaker.

Also, in the following architecture, machine learning algorithms [1, 2] are used in the analysis of dermatological diseases and classification of phono-cardiographic signals.

2. PCG Based Ayurvedic Analysis

Signal based processing is one of the important research areas. Signals may be of different types like image, sound and electromagnetic. The signals generated from living bodies are classified as biomedical signals and can be processed for a variety of applications including health state estimation [3–10]. The signals may be obtained from different physiological systems in the human body such as the circulatory, muscular and nervous systems. As the heart is an important organ in the human body, the signals recorded around it provide significant information about the functioning of the body. The audio signals in the form of heart sounds, usually called phonocardiogram (PCG) signals, contain enough information about the working of the heart in particular and working of the body in general. Because of the complexity involved in the analysis of the heart sounds, the interpretation of the experts may vary depending upon their skill and expertise. Hence, there is a need for automating the process of diagnosis and decision-making using machine learning algorithms. The researchers have focussed on how Ayurveda, an ancient Indian medical science understands and visualizes the human body. The ayurvedic body is conceptualised as being composed of five constituent parts (*Panchamahabhutas:* space, air, fire, water and earth) [11–13] and a hierarchical relationship exists between these elements [14, 15] that is represented in Fig. 2.

Ayurveda defines a definite relationship between the functioning of the heart and three fundamental quantities (*Vata, Pitta, Kapha*), also known as

Fig. 2. Representation of *Panchamahabhuta* and *Tridosha* (modified from [15]).

Tridosha that shows the overall functioning of living bodies. In DSP Lab, Jammu University, India, research is being carried out for designing a system using convolutional neural networks for estimating the relationship between the PCG signals and *Tridoshas*. This automated system could be used for designing medical expert systems for predicting health related problems. The hypothesis of their ongoing research work is that the *Tridoshas* can be derived from the phonocardiographic (PCG) signals using correlation, regression, and neural network based function estimation. In their work, the most dominant *dosha* has been estimated from PCG signals using CNN. Further comparison of the results have been done using multivariate quadratic modelling (MQM), which is a regression based estimation technique.

In the first phase of developing the system in the DSP lab, a recording of phono-cardiographic signals was carried out followed by their analysis using machine learning algorithms.

Recording the Signal

The heart sounds or PCG signals of 30 subjects have been recorded in an acoustically tested room. The subjects have been instructed to enter the room one at a time and the process of 'Wrist Pulse Examination' by an ayurvedic physician has been carried out in which the subject's radial artery has been examined for *dosha* imbalance. In male subjects, the right hand's pulse and in female subjects, the left hand's pulse has been examined. The subject has been instructed not to eat anything for atleast two hours prior to the examination. The examination has been performed with the subject sitting in a stable and upright position. After the wrist pulse examination (Fig. 3), the heart pulse sounds of the subject have been recorded using Jabes-Advanced Bio Electronic Stethoscope (Fig. 4). The acquisition device has been interfaced to an amplifier. The amplifier that has been used is SSB-45EM Ahuja Radio. The amplifier has been interfaced with a computer application to record the signal at 1000 Hz Signal frequency for 3 minutes.

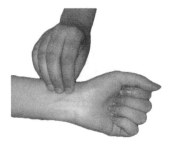

Fig. 3. Wrist pulse examination by ayurvedic physician.

Fig. 4. Recording set-up for PCG signals.

Machine Learning Based Analysis of PCG Signals

The PCG signal dataset obtained was converted into a time-frequency representation and represented using a scalogram (absolute value of the continuous wavelet transform coefficients of the signal) which were further converted into RGB images of size 224 × 224 × 3 that were fed as input to the convolutional neural network. Pre-trained GoogLeNet [26] CNN model was used to perform the classification of the PCG signals into three classes: *Vata*, *Pitta* and *Kapha*. An example of the zoomed view of recorded PCG signals (amplitude vs. time) belonging to the class *Vata* (Fig. 5).

Investigation of machine learning based analysis of their work shows that the accuracy of PCG signal classification using CNN varies according to the parameters specified during training like learning rate, number of epochs, batch size and activation functions. The accuracy reaches a maximum of approximately 86% after training the CNN for 6 epochs. Work is being

Fig. 5. PCG signal of class *Vata*.

carried out for increasing the accuracy of the experiment by varying different parameters.

3. Automated Wound Healing

Wound dressings play an important role in managing/healing of wounds. A small cut or scratch poses a threat to vulnerable infections in humans with a weak immune response. The skin has a regenerative property that supports fast healing. In case of burn injuries and deep cuts, the regenerative property of the skin is not strong enough and it fails to supply sufficient nutrients and other supportive components that accelerate the repair process. A suitable environment for healing like sufficient oxygen/moisture level in the wounds and protection from bacteria are necessary factors for repairing the damaged tissues/cells. A smart bandage is an artificial intelligence system that senses the need of the wound and supplies drugs to it accordingly to the per pre fed pattern. Smart bandages are a boon to the people in remote rural areas or in the case of the pandemic. Some wounds require regular visits to the doctor to examine the wound status. The smart bandage technology is based on detecting the change in pH, oxygen/moisture level, and temperature of the wound as these four components play a significant role in healing the wound. The section below provides the contribution of researchers worldwide towards smart bandage technology. Lo et al. has designed a microfluidic oxygen bandage as oxygen is the main factor that helps in healing a wound. Oxygen bandages are easy to fabricate and are an economical method for restoring tissue oxygenation. A smart bandage based on microfluidic technology is fabricated for diffusing oxygen into the wound. With this microfluidic bandage, 70% oxygen supply is diffused across the wound in 7 days that results in collagen maturation in the wound area. Oxygen concentrations are delivered at a fast speed from 0 to 100% in 60 seconds [16]. Booteng et al. has reviewed the different dressings suitable for different wounds. The dressings requirement differs from wound to wound. Traditional and modern systems of wound dressings are useful but have certain shortcomings as they are unable to replace lost tissue in cases of burn injuries. Smart polymers are either composed of natural or semi synthetic material, used as a replacement for the dead tissues. These show a normal physiologic response during wound healing. Biological dressings use biomaterial for improved healing at a fast rate. Collagen bio dressing stimulates the formation of fibroblasts as collagen plays an important role in wound healing. Controlled drug delivery to the wound is another effective method of healing it. Fluid handling properties, fluid affinity/retention, tensile strength, elasticity and bio adhesive strength are the factors affecting wound dressings. The review of literature suggests that it is difficult to find a dressing that is meant for every type of wound [17]. Farroqui et al. reported that wounds with pH variation are prone to infections. Physical inspection of wounds is a necessity for smooth healing. It requires frequent visits to hospitals which is a costly affair besides being time consuming. Researchers have reported a smart wearable bandage that is able to detect signs of infection at an early stage. Pressure sensors sound an alarm to the patient that helps in avoiding pressure ulcers. It alerts the patients to changes in the dressing. These help doctors to monitor the wound by sitting at distant place. The data is sent through the internet to a doctor at a distant place. There are microcontrollers, LEDs for displaying the status of wound [18]. Rahimi et al. reported another pH based bandage for monitoring the wound healing process to avoid the spread of infection. The monitoring bandage system consists of two parts disposable and reusable. The disposable part consists of flexible pH sensors placed in contact with the wound [19]. Yoshida et al. has reported stimuli responsive hydrogels that may help in controlling the flow rate, buoyancy force and helps in the realization of complex movements [20]. Mostafalu et al. also reported a smart microfluidic bandage composed of disposable and reusable electronic parts (Fig. 6). The smart bandage comprises of temperature and pH sensors and a micro heater for heating the hydrogel patch to drain the thermoresponsive drugs to the wound. The temperature sensors provide information about the inflammation of the wound and also helps in adjusting the temperature. A wireless electronics module is attached to check the change in data from the sensors and initiates a control signal to trigger the drug release process. This smart bandage does not just monitor the change in pH but also release of the drug and intimation regarding the change to the doctor. The bandage comprises of a flexible heater that is controlled by the microcontroller in the electronic module. The bandage is cost effective with low cost sensing and heating modules so that they can

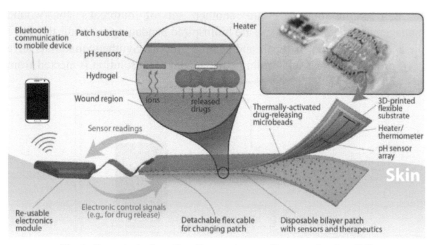

Fig. 6. Conceptual schematics of temperature sensitive smart bandage [21].

be easily disposed of. The microcontroller communicates with the external devices with a cost effective Bluetooth device [21].

Bagherifard et al. developed a hydrogel dermal patch encapsulated with a thermoresponsive drug. It is attached to the electronic heater control circuitry. The drug delivery is initiated by adjusting the thermo sensitive hydrogel layer [22]. Pressure applied to a live tissue for an extended period of time may result in blood loss and may even cause cell death if left unattended. This leads to pressure ulcers. Swisher et al. reported a smart device for the detection of pressure ulcers with impedance spectroscopy. It consists of an electronic sensing instrument that detects the change in impedance level with a multiplexed electrode array. A flexible bandage array and rigid pcb with gold plated electrodes was developed for sensing the impedance change [23]. Fan et al. has reported a smart wound dressing capable of maintaining sufficient moisture and antibacterial properties for keeping the pathogens away and to pace the healing process. Ag5G1 composite hydrogel was taken with 5:1 Ag to graphene mass ratio. It was investigated that the composite has sufficient antibacterial properties against Gram-positive *Staphylococcus aureus* and Gram-negative *Escherichia coli* [24]. Uncontrolled bleeding in a wound can be reduced by using a hemostatic agent. Commonly used hemostatic agents are either derived from animal or human products and they increase the risk of carrying infections to a healthy person sometimes. Hardy et al. reported a microelectronic bandage (Fig. 7) that helps in fast clotting by applying low voltage. The low voltage applied helps in fast fibrin formation without excessive heating of the wound site. It was analyzed that a voltage below 15 V produces a current, not more than 0.03 A. A voltage between 10–30 V applied to a simulated tissue produces a visible black charring on the wound. Voltages less than 9 V do not produce any visible ill impact [25].

Self-healing materials are another set of outbreaks for wound management technology. These are bioinspired bandaging materials that have the characteristic to self-repair and healing process. Synthetic materials have been produced to test self-healing mechanisms. The solution is ejected from

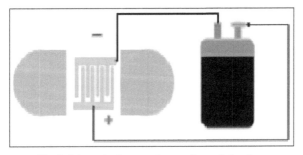

Fig. 7. Schematic diagram of smart electronic bandage.

the vascular structures into the cracked area, mixes and hardens to form the new healed surface [26].

Traditional bandages have a lot more disadvantages. The bacterial infection can easily thrive and contamination may increase. Maintaining an appropriate moisture level is difficult as due to high permeability wounds may dehydrate. Changing dressings may cause mechanical tearing of the healed tissue. It is really a discomfort for the patients especially for kids while removing or changing dressings. The fibers in the dressing get stuck to the wound. All these shortcomings are being taken care in smart wound bandaging techniques. Smart bandages sense the state of the wound, supply the drug accordingly wherever needed and monitor the healing process without actually visiting the doctor [27]. It was concluded from the traditional and modern bandaging techniques that a blend of ayurvedic science and microfluidic technology may help in better healing of wounds. Flexible smart bandages (Fig. 8) are the need of the hour. The bandages should be fabricated immediately as per the need of the patients and drugs could be encapsulated in the chamber for immediate benefit. Arc Sign Printer is a cheap and cost effective solution for flexible printing as compared to PDMS that requires hours for fabricating the bandage structure.

The proposed structure for smart bandaging (Fig. 9) consists of drug chambers, microfluidic channels, mixers, active area, status bar, sink, control signals, and microcontroller, small dc power supply.

A Nichrome wire is used to control the flow of drugs into the microfluidic channels. The mixers are used to ensure the uniform distribution of each drug to the active area of the bandage. There are two wires in touch with the active area to supply sufficient amount of heat to the wound. pH, oxygen sensors are placed in direct contact with the wound to ensure proper supply of drugs to it. As soon as change in pH and oxygen is detected, signals are sent to the microcontroller. It switches on the circuit and sends an electric signal to nichrome wire. The heat generated shrinks the drug chambers and drugs are pushed out of the closed drug chambers. The quantity of drug release can be controlled by controlling the heat of the nichrome wires. Mixers are used

Fig. 8. Flexible microfluidic structure fabricated with transparent sheet and copper connections.

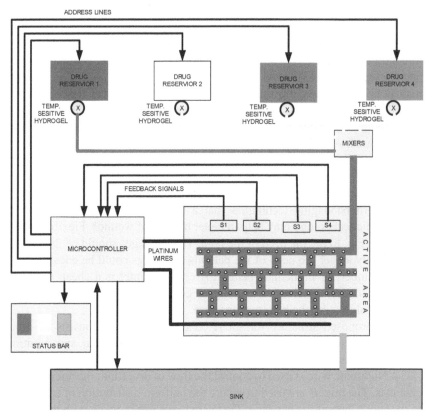

Fig. 9. Schematics of the working module of microfluidic smart bandage.

to mix the different drugs drained out of the drug chambers. As per wound need, if only one drug is required, the rest of the drug chambers are kept in off condition. In that case only one drug passes through the mixer and mixing does not take place. Drugs are diffused to the active area of the wound and the status of wound healing is displayed on the status bar. The supply of drug is controlled by change in the oxygen and pH levels of the wound. The system sounds an alarm if either the drug chambers goes empty or the sink gets full. The smart bandage machine may be programmed to release the specific drug as per the signal sent to a particular drug chamber. The bandage may be a boon to the people in remote areas or in the battle field where smart techniques for managing the wounds is a prime necessity. The bandages can be printed within seconds and can be filled with desired drugs [28, 29].

4. Speech Based Psychological Analysis

The glottal speech analysis of a person helps in determining the emotional expression and its relation to the overall state of the speaker. The creation of a framework for performing real-time analysis of speech for detecting emotional states and assessing the quality of speech has been receiving considerable attention. A system named Pepper robot has been designed as a practical application of speech emotion recognition [30]. Although, it has not been used for any medical applications or clinical diagnostics. There is a need for developing a system which integrates speech emotion recognition with the emerging communication technologies in order to come up with modern healthcare systems catering to the needs of rural India. Such systems may help in keeping a track of the health of an individual by a medical expert or any other user designated for the purpose.

The system described in [31] uses next generation networks for transmission of data and cloud based system for data analysis and storage. The system can be monitored using mobile applications by the users.

For data collection, wearable devices with sensors are provided which monitor the user's body temperature, emotions through user speech, heart rate and more. The data is transmitted using the network for data analysis and interpretation. Figure 10 shows the fractals for the unaspirated. It has been proposed to integrate different machine learning modules to deliver smart healthcare in rural areas. When an individual is under emotional stress, there is an aberration in the amount of tension applied in closing, i.e., adduction as well as opening, i.e., abduction of the vocal cords. This aberration can cause significant changes in the volume-velocity air profile through the vocal cords as well as fluctuations in the fundamental frequency. Information on the movement of the vocal cords can only be measured through the evaluation of glottal waveforms. Cepstral features have also been used effectively for emotional analysis of speech. Prosodic, glottal, cepstral and vocal track features come under physiological phenomenon. Whereas another feature which has been used is Teager energy operator which comes under perceptual phenomena. These features may also be classified based on the linearity of airflow in speech production into linear and non-linear features. The work related to the use of speech processing in healthcare reflects that prosodic characteristics have proved to be most effective. Parameters including fundamental frequency, formants, jitter, shimmer, intensity of the speech signal, and speech rate have been studied. Other commonly used speech parameters were Mel Frequency Cepstral Coefficients (MFCCs) and power spectral density (PSD). Due to computational complexity of the above discussed methods, fractal analysis (Fig. 11) has been used in [32] for speech emotion analysis. It was reported that fractal dimensions as well as fractal

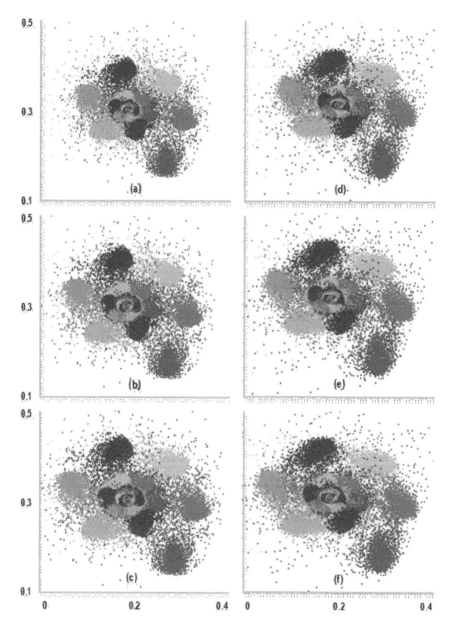

Fig. 10. Convergent fractal functions for unaspirated phonemes. Column 1 (a–c) shows /a:pa:/ in neutral, happy and angry speech respectively. Column 2 (d–f) shows /a:ba:/ in neutral, happy and angry speech respectively.

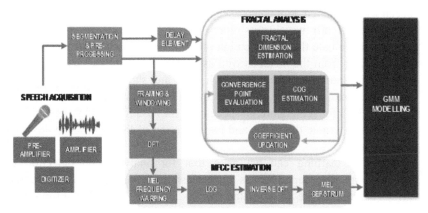

Fig. 11. Block diagram of speech classification.

loop areas were found to be different for different emotional speeches. This was used in a GMM-based model to predict the emotional state of the speaker. Using this method involving very less computation, the emotional state of rural patients can be monitored easily through their speech.

Many mathematical models are being used for efficient emotion classification such as Gaussian Mixture Models (GMM), Support Vector Machines (SVM), harmonic plus noise model (HNM), decision tree methods, Multilayer Perceptron neural networks (MLP) and Hierarchical Fuzzy Signature (HFS) classifier. The accuracy of classification can be increased by using hybrid classifiers like a combination of GMM and SVM models. Most of the present applications require the systems to have a self-learning capacity. This makes neural networks the most preferred choice for classification instead of mathematical models. Neural network models are efficient, have an adaptive mechanism, the ability to generalize and are much faster due to a parallel computational structure. In recent years, Artificial Neural Networks (ANN), Deep Neural Networks (DNN), Recurrent Neural Networks (RNN), Convolutional Neural Networks (CNN) have exhibited an outstanding performance in extracting discriminative features for emotion recognition. Compared with other hand-crafted features, different types of neural networks are capable of extracting hierarchical feature representations for any specific task such as emotion recognition given a large amount of training data by the use of supervised learning. Hence, apart from use in monitoring essential healthcare parameters, neural networks have proved to be the best classifiers for the task of emotion recognition from speech as well. As discussed in [33] various speech parameters including mean pitch, median pitch, standard deviation, minimum pitch, maximum pitch, jitter and shimmer were analyzed. The extracted feature vectors were used to represent each sentence of schizophrenic and healthy speech. The mean vector for both categories of speech was estimated. K-means based centre of gravity was used to divide the feature vectors into two clusters representative of both categories of speech. In order to evaluate the identification rate, the Mahalanobis distance of each feature vector from the mean as well as centre of gravity was estimated and identification rate was calculated. Pitch, jitter and shimmer were found to be most useful parameters for identification of schizophrenic individual with a rate of 77.35% for schizophrenic speech and 84.70% for healthy speech. This could help in remote diagnosis and identification of the disorder in the rural population.

Machine learning can act as a tool for analysis and detection of psychological disorders such as hypertension, depression, and anxiety if various parameters are accumulated and studied over a duration of time. This could in future also help in developing a correlation between various chronic disorders and the emotional state of the user. It would also help in remote and efficient monitoring of rural patients. It may also be able to predict the pattern of occurrence of some disorders and hence act as a preventive measure. Certain disorders like depression cannot be diagnosed by biomedical methods but such smart systems may help in their early diagnosis and hence provide timely treatment to patients especially in rural India.

5. Dermatological Analysis

Telemedicine applicability in the field of healthcare has gained significant importance during the COVID-19 pandemic by providing expert consultation to remote areas without in-person consultations, reducing incurred healthcare costs. Tele dermatology is appropriate for this kind of care system and is defined as the use of communication technology for dispensing dermatological services [34]. Tele dermatology has different formats: a real-time, live-interactive technology, automated smartphone apps, and a store-and-forward technology. Live-interactive tele dermatology, uses live video conferencing between the patient and dermatologist, whereas, in store-and-forward tele dermatology, dermascopic images are acquired by the inexperienced medical staff at rural clinics which are sent to expert dermatologists situated in urban catheterization labs [35]. In contrast, automated smart phone apps employ machine learning (ML) algorithms to find out on the spot the probability of malignancy without consultation from the dermatologist. Tele dermatology and personal monitoring applications combined with ML algorithms have great potential these days due to their portability and accessibility.

Machine Learning algorithms are useful for diagnosis of various types of dermatological disease images such melanoma and psoriasis captured through various sources like smartphones, DSLRs, dermoscopic, histopathology and datasets for large scale epidemiology research. Several researchers have done experiments showing the effectiveness of machine learning algorithms in melanoma and non-melanoma skin cancer diagnosis. Melanoma diagnosis using machine learning algorithms was first employed by Nasr–Esfahani et al. [36] reporting 0.81 sensitivity and 0.80 specificity for their system. Further in 2017, research was conducted by Stanford University using convolutional neural networks on dermatological clinical images for their cancer classification [37]. The limitation of their research was that large datasets were needed for carrying out the research. This was overcome by another research conducted by Fujisawa et al. [38], where the system employing deep convolutional neural networks was efficiently used for dermatological cancer classification using a small clinical images dataset. The overall accuracy achieved by the system was 76.5%. Further, their system reported 96.3% sensitivity and 89.5% specificity. Han et al. [39] also developed a system using machine learning algorithms to classify skin disease images, reporting an average sensitivity and specificity of 85.1% and 81.3%, respectively.

Brinker et al. [40] reported supremacy of automated image classification of melanoma using machine learning techniques in comparison to traditional methods of diagnosis by dermatologists.

Investigations on the application of the Gaussian mixture model (GMM) based on the analysis and classification of skin diseases from their visual images using a Mahalanobis distance measure were carried out by Chaahat et al. [41]. In their work depicted in Fig. 12, the GMM has been preferred over the convolution neural network (CNN) because of limited resources available within the mobile device. Gray-level co-occurrence matrix parameters contrast, correlation, energy, and homogeneity derived from skin images used as the input data for the algorithm. The analysis of their results showed that their methodology was able to predict the classification of skin diseases with satisfactory efficiency.

In their investigations, for instance in Fig. 13, investigations were carried out for Acne-Blackhead in which the red component of the GMM modeled feature vector set showed important peaks at 0.46, 0.34, 0.39 for contrast, at 0.09, 0.24, 0.14 for correlation, at 0.29, 0.35, 0.33 for energy, and at 0.78, 0.83, 0.81 for homogeneity. Its green component showed important peaks at 0.47, 0.30, 0.37 for contrast, at 0.10, 0.22, 0.20 for correlation, at 0.29, 0.41, 0.34 for energy, and at 0.78, 0.85, 0.82 for homogeneity. Lastly, the blue component of this disease showed important peaks at 0.30, 0.40, 0.19 for contrast, at 0.27, 0.18, 0.31 for correlation, at 0.39, 0.31, 0.54 for energy, and at 0.85, 0.80, 0.90 for homogeneity.

The comparisons of their classifications based on Euclidean and Mahalanobis distance measures of different diseases with respect to normal skin using 72-dimensional feature vectors are visually shown in

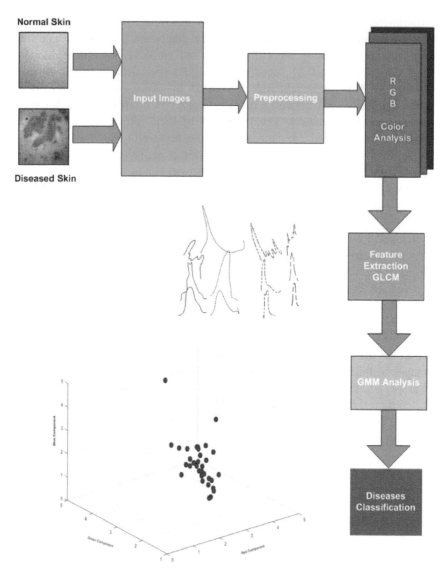

Fig. 12. Proposed methodology. GLCM parameters of normal and diseased skin are calculated after resizing each color component of the images. Training of GMM is carried out and classification is performed using Euclidean/Mahalanobis distance measure approaches for classification [T].

Fig. 14 to Fig. 17. It was observed that different diseases occupy different spatial positions in their scatter plots. Mahalanobis based scatter plots (Fig. 15) showed better results as dissimilar diseases got relatively more scattered as compared to that of Euclidean based scatter plots (Fig. 14). Further,

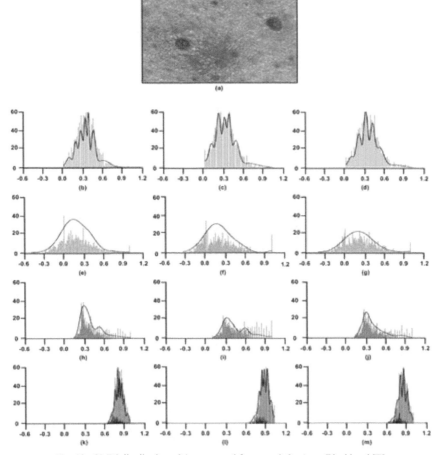

Fig. 13. GMM distribution of the proposed framework for Acne-Blackhead [T].

instances of same disease (e.g., Melanoma in Fig. 17) gave close grouping as compared to Euclidean based scatter plots in Fig. 16 in their investigations.

Also, Chaahat et al. [42] have done research on automated image enhancement, followed by a CNN based skin lesions diagnosis for applications in resource deficient environments like rural areas. Investigations showed that the machine based algorithm was able to enhance the classification accuracy from 87.40% to 95.85% when Genetic algorithm (GA) enhanced images were used for diagnosis. The GA based enhancement was able to improve the blurred images to a satisfactory level. The results of their machine learning based classification of dermatological diseases is depicted in Fig. 18.

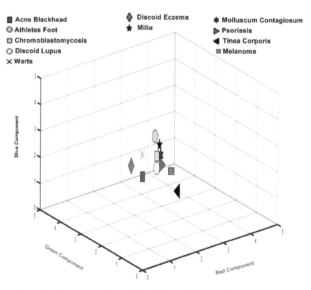

Fig. 14. Three-dimensional scatter plot of Euclidean distance for 11 skin diseases taken for investigation [T].

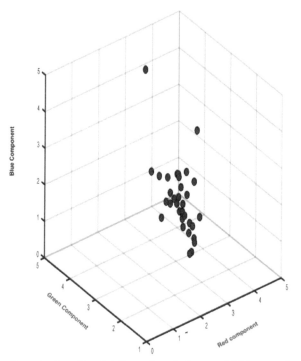

Fig. 15. Three-dimensional scatter plot for various instances of Melanoma disease taken for investigation using Euclidean distance measure approach [T].

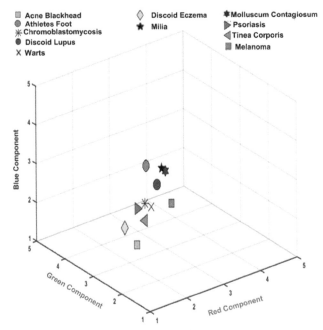

Fig. 16. Three-dimensional scatter plot of Mahalanobis distance for 11 skin diseases taken for investigation [T].

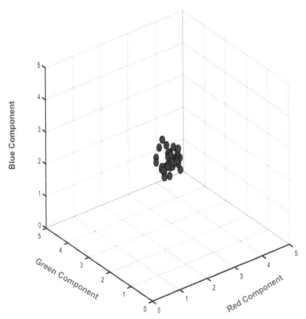

Fig. 17. Three-dimensional scatter plot for various instances of Melanoma disease taken for investigation using Mahalanobis distance approach [T].

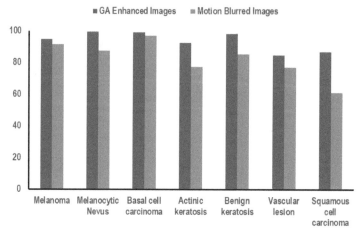

Fig. 18. Classification accuracy of the individual classes (%) [42].

6. Smart Hot and Cool Automated Relief Pack

Hot and Cold therapy have been used since long for treating muscle injuries, ankle sprains, cramps, spasms, bruising, inflammation muscle knots and improving circulation of blood to the injured tissues. Heating causes the blood vessels to expand and cooling causes the vessels to contract providing pain relief. Our brain is unable to detect temperature change and pain at the same time and on applying hot and cold therapy, the brain senses the temperature variation thus relieving us from pain temporarily. Hot or warm packs are applied in areas of neck muscle pain, muscle spasms and joint stiffness. Heat improves the muscle flexibility, eases its tension and helps in speeding up blood circulation. Cold packs are used in case of arthritis, acute injuries, tendonitis or bursitis. It reduces inflammation, slows metabolism and reduces skin temperature. The reduction in the metabolism in the injured area helps in decreasing the tissue damage. Ice also slows the transmission of pain signals to the brain. Hot and cold therapy differs from injury to injury. Some injuries require only hot therapy whereas some need only cold therapy. It's always advised to visit the therapist before trying any therapy at home. This practice has been used since long. Initially people used a cloth soaked in warm and cool water separately for treatment of such injuries. Nowadays, doctors prescribe the same procedure for certain injuries at home but everyone can't afford to visit the physiotherapist for such specific injuries. It's also tiresome to perform the therapy at home for a long period. Hot and cold packs are available now a days to be used at home for treating specific injuries. However, the therapies requiring hot and cold at the same time after regular time intervals are not available in the market and the same effect is provided by the Peltier element. The Peltier effect is a temperature difference created by applying a voltage between two electrodes connected to a sample of semiconductor material [43, 44]. This phenomenon can be useful when it is necessary to transfer heat from one medium to another on a small scale.

A further, invention conducted by researchers in published patent [201811025275 A] provided a hot and cold relief system consisting of six blocks. The research was carried out in DSP Lab, Jammu University. A mobile app, wifi module, processor, interface, temperature sensor device and power supply of 3.3 V, 5 V and 12 V were utilized. The interface worked on 5 V or 12 V, the processor required 5 V and wifi modules ran on 3.3 V. Further mobile application used wifi signals and wifi module to run hot and cold cycles as desired by the user. The designed processor transmitted a signal to the temperature sensors through the interface unit to program the Peltier based hot and cold pack. A liquid gel based pack heated up and cooled down as per the cycle of the Peltier element used. The switching of the Peltier element between the hot and cold cycle was programmed. Their proposed block diagram representations are shown in Fig. 19.

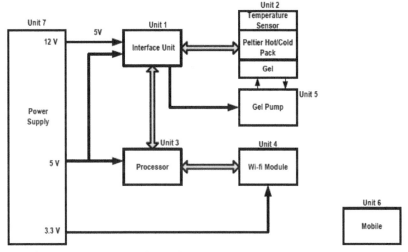

Fig. 19. Automated Peltier based hot/cold muscle relaxer [201811025275 A].

References

[1] Machine Learning Algorithms. 2018. Optimization Techniques and Applications with Examples, pp. 199–226, Oct. 2018.

[2] Ayyadevara, V. K. 2018. Artificial Neural Network. Pro Machine Learning Algorithms, pp. 135–165.

[3] Najarian, K. and Splinter, R. 2016. Signals and biomedical signal processing. Biomedical Signal and Image Processing, pp. 3–14, Apr. 2016.

[4] Farina, D. and Holobar, A. 2016. Characterization of human motor units from surface EMG decomposition. In Proceedings of IEEE 104(2): 353–373.

[5] Pal, N. R., Chuang, C.-Y., Ko, L.-W., Chao, C.-F., Jung, T.-P. et al. 2008. EEG-based subject- and session-independent drowsiness detection: an unsupervised approach. EURASIP Journal on Advances in Signal Processing, vol. 2008, no. 1, Nov. 2008.

[6] Sravani, K. and Rao, R. 2017. High throughput and high capacity asynchronous pipeline using hybrid logic. 2017 International Conference on Innovations in Electronics, Signal Processing and Communication (IESC), Apr. 2017.

[7] Baumert, M., Porta, A. and Cichocki, A. 2016. Biomedical signal processing: from a conceptual framework to clinical applications [Scanning the Issue]. Proceedings of the IEEE 104(2): 220–222, Feb. 2016.

[8] Sharma, A. and Abrol, P. 2013. Eye gaze techniques for human computer interaction: a research survey. International Journal of Computer Applications 71(9): 18–25, Jun. 2013.

[9] Sharma, A. and Abrol, P. 2016. Direction estimation model for gaze controlled systems. Journal of Eye Movement Research vol. 9, no. 6, Sep. 2016.

[10] Nandha, R. and Singh, H. 2013. Amalgamation of ayurveda with allopathy: A synergistic approach for healthy society. International Journal of Green Pharmacy 7(3): 173.

[11] Upadhyay, D. and Dwibedy, B. K. 2017. A review on ayurvedic practice through sadapadartha theory. International Journal of Research in Ayurveda & Pharmacy 8(3): 31–35, Jul. 2017.

[12] Metri, K., Bhargav, H., Ramarao, N., Rizzo-Sierra, Rr. Basavakatti et al. 2012. OA01.25. The first direct experimental evidence correlating ayurveda based tridosha prakriti, with western constitutional psychology somatotypes. Ancient Science of Life 32(5): 25, 2012.

[13] Clayton, M. 2001. Sharma, Shiv Kumar. Oxford Music Online.

[14] Sahi, P., Abrol, P., Kumari, S. and Lehana, P. K. 2018. Origin of Object Oriented Computing Models for System Development in Indian scriptures, presented at the Third World Congress of Vedic Sciences, Pune, Jan 2018.

[15] GEO LYONG LEE. 2016. A study on the concept of agni in Āyurveda. Journal of South Asian Studies 22(2): 103–139, Oct. 2016.

[16] Lo, J. F., Brennan, M., Merchant, Z., Chen, L., Guo, S. et al. 2013. Microfluidic wound bandage: Localized oxygen modulation of collagen maturation. Wound Repair and Regeneration 21(2): 226–234, Feb. 2013.

[17] Boateng, J. S., Matthews, K. H., Stevens, H. N. E. and Eccleston, G. M. 2008. Wound healing dressings and drug delivery systems: a review. Journal of Pharmaceutical Sciences 97(8): 2892–2923, Aug. 2008.

[18] Farooqui, M. F. and Shamim, A. 2016. Low cost inkjet printed smart bandage for wireless monitoring of chronic wounds. 2016 IEEE MTT-S International Microwave Symposium (IMS), May 2016.

[19] Rahimi, R., Brener, U., Ochoa, M. and Ziaie, B. 2017. Flexible and transparent pH monitoring system with NFC communication for wound monitoring applications. 2017 IEEE 30th International Conference on Micro Electro Mechanical Systems (MEMS), Jan. 2017.

[20] Yoshida, K., Nakajima, S., Kawano, R. and Onoe, H. 2017. Stimuli-responsive hydrogel microsprings for multiple and complex actuation. 2017 IEEE 30th International Conference on Micro Electro Mechanical Systems (MEMS), Jan. 2017.

[21] Mostafalu, Tamayol, A., Rahimi, R., Ochoa, M., Khalilpour, A. et al. 2018. Smart bandages: smart bandage for monitoring and treatment of chronic wounds (Small 33/2018). Small 14(33): 1870150, Aug. 2018.

[22] Bagherifard, S., Tamayol, A., Mostafalu, P., Akbari, M., Comotto, M. et al. 2015. Dermal patch with integrated flexible heater for on demand drug delivery. Advanced Healthcare Materials 5(1): 175–184, Oct. 2015.

[23] Swisher, S. L., Lin, M. C., Liao, A., Leeflang, E. J., Khan, Y. et al. 2015. Impedance sensing device enables early detection of pressure ulcers *in vivo*. Nature Communications vol. 6, no. 1, Mar. 2015.

[24] Fan, Z., Liu, B., Wang, J., Zhang, S., Lin, Q. et al. 2014. A novel wound dressing based on Ag/graphene polymer hydrogel: effectively kill bacteria and accelerate wound healing. Advanced Functional Materials 24(25): 3933–3943, Mar. 2014.

[25] Hardy, E. T., Wang, Y. J., Iyer, S., Mannino, R. G., Sakurai, Y. et al. 2018. Interdigitated microelectronic bandage augments hemostasis and clot formation at low applied voltage *in vitro* and *in vivo*. Lab on a Chip 18(19): 2985–2993.

[26] Selimović and Khademhosseini, A. 2012. Research highlights. Lab Chip 12(1): 17–19.

[27] COMPRESSION BANDAGING 1.1. Bandage Features and Function. Wound Healing, pp. 439–440, Aug. 2005.

[28] P. R. et al. 2021. Design and simulation of microfluidic smart bandage using Comsol multiphysics. Turkish Journal of Computer and Mathematics Education (TURCOMAT) 12(5): 1650–1662, Apr. 2021.

[29] Rajput, P. and Lehana, P. K. 2020. Fabrication of smart bandage structure. International Journal of Scientific and Technical Advancements, pp. 111–114.

[30] Farooqui, M. F. and Shamim, A. 2016. Low cost inkjet printed smart bandage for wireless monitoring of chronic wounds. Scientific Reports vol. 6, no. 1, Jun. 2016.

[31] Lee, D.-J., Park, M. and Lee, J.-H. 2015. Height adjustable Multi-legged Giant Yardwalker for variable presence. 2015 IEEE International Conference on Advanced Intelligent Mechatronics (AIM), Jul. 2015.

[32] Abrol, A., Kapoor, N. and Lehana, P. K. 2020. Speech recognition based emotion-aware health system. International Journal of Scientific and Technical Advancements.

[33] Abrol, A., Kapoor, N. and Lehana, P. K. 2021. Fractal-based speech analysis for emotional content estimation. Circuits, Systems, and Signal Processing, pp. 1–22.

[34] Abrol, A., Kapoor, N. and Lehana, P. K. 2021. Parametric analysis of schizophrenic speech. International Journal of Scientific and Technical Advancements.

[35] Hadeler, E., Gitlow, H. and Nouri, K. 2020. Definitions, survey methods, and findings of patient satisfaction studies in teledermatology: a systematic review. Archives of Dermatological Research 313(4): 205–215, Jul. 2020.

[36] Sonar, P. 2021. High performance organic semiconductors for photonics and electronics. Video Proceedings of Advanced Materials 2(2): Article ID 2021–02110–Article ID 2021–02110.

[37] Nasr-Esfahani, E., Samavi, S., Karimi, N., Soroushmehr, S. M. R., Jafari, M. H. et al. 2016. Melanoma detection by analysis of clinical images using convolutional neural network. 2016 38th Annual International Conference of the IEEE Engineering in Medicine and Biology Society (EMBC), Aug. 2016.

[38] Esteva, A., Kuprel, B., Novoa, R. A., Ko, J., Swetter, S. M. et al. 2017. Dermatologist-level classification of skin cancer with deep neural networks. Nature 542(7639): 115–118, Jan. 2017.

[39] Fujisawa, Y., Otomo, Y., Ogata, Y., Nakamura, Y., Fujita, R. et al. 2018. Deep-learning-based, computer-aided classifier developed with a small dataset of clinical images surpasses board-certified dermatologists in skin tumour diagnosis. British Journal of Dermatology 180(2): 373–381, Sep. 2018.

[40] Han, S. S., Kim, M. S., Lim, W., Park, G. H., Park, I. et al. 2018. Classification of the clinical images for benign and malignant cutaneous tumors using a deep learning algorithm. Journal of Investigative Dermatology 138(7): 1529–1538, Jul. 2018.

[41] Brinker, T. J., Hekler, A., Enk, A. H., Berking, C., Haferkamp, S. et al. 2019. Deep neural networks are superior to dermatologists in melanoma image classification. European Journal of Cancer 119: 11–17, Sep. 2019.

[42] Gupta, C., Gondhi, N. K. and Lehana, P. K. 2019. Analysis and identification of dermatological diseases using Gaussian mixture modeling. IEEE Access 7: 99407–99427.

[43] Chaahat, N. Kumar Gondhi and Kumar Lehana, P. 2021. An evolutionary approach for the enhancement of dermatological images and their classification using deep learning models. Journal of Healthcare Engineering 2021: 1–13, Jul. 2021.

[44] Nesarajah, M. and Frey, G. 2016. Thermoelectric power generation: Peltier element versus thermoelectric generator. IECON 2016 - 42nd Annual Conference of the IEEE Industrial Electronics Society, Oct. 2016.

[45] Boehm, R. F. 2000. The CRC Handbook of Thermal Engineering. Ed. F. Kreith (CRC, Boca Raton), pp. 3-1–3-14.

CHAPTER 6
Swarm Intelligence-based Framework for Image Segmentation of Knee MRI Images for Detection of Bone Cancer

Sujatha Jamuna Anand,[1] *C Tamilselvi,*[2] *Dahlia Sam,*[2]
C Kamatchi,[3] *Nibedita Dey*[4] and *K Sujatha*[5,*]

1. Introduction of Bone Cancer

The human body is categorized into connective, sensory, epithelial and muscle tissues. The connective tissue is categorized as fitting connective and specific connective. Specific connective tissues consists of blood vessels and has a hard consistency. The rigidity of the bone is due to minerals and collagen strands. The human skeleton system consists of two hundred and six bones. The difference in the intensity level of the osteon indicates the minerals present in it [1].

---

[1] Principal, Loyola Institute of Technology, Chennai, 600 123.
[2] Assistant Professor, Dept. of IT, Dr. MGR Educational and Research Institute. Chennai.
[3] Associate Professor, Dept. of Biotechnology, The Oxford College of Science, Chennai.
[4] Assistant Professor, Dept. of Biotechnology, Saveetha Univeristy, Chennai.
[5] Professor, EEE Dept., Dr. MGR Educational and Research institute, Chennai.
Emails: sujja13@gmail.com; tamilselvi.cse@drmgrdu.ac.in; dahliasam@drmgrdu.ac.in; ckamatchi@gmail.com; nibeditadey.sse@saveetha.com
* Corresponding author: sujathak73586@gmail.com

Cancer Formation in Bones

Bone growth is an uncommon disease that creates a disturbance in the bone called bone tumour. Malignant growth of the bone is known as Osteosarcoma. At the point when a bone malignancy starts to develop, the disease cells increase and begin to demolish the bone. The uncontrolled bone growth in the regions of knee, wrist, shoulder and pelvis causes bone tumour. Thirty different types of bone tumours are identified. The most widely recognized type is the osteosarcoma. There could be a development of a stage state, which spreads out with other parts or to organs known as metastasis. The various phases of bone tumor are

Phase I: the malignancy confined to that bone structure alone

Phase II: identical to Phase I, but it is a powerful one

Phase III: tumours exist in different regions of the same bone

Phase IV: growth has spread to other organs in the body.

Estimated cancer cases and deaths across USA is tabulated in Table 1 and Estimated cancer cases and deaths across globe is shown in Table 2.

The traditional approaches of bone cross section were performed using microscopic image analysis which are acquired by using micro radiography, circularly polarized light scans, transmitted light scans, plain polarized light scans and different levels of mineralization in bone cross section variations were acquired in grey level intensity. However, the examination of microscopic images leads to blunders because of structural and temporal variations of objects in raw images and also needs well trained operators for scrutiny. To identify the pathological changes, CTs, SPECT and PET which provide the anatomical structure, the functional information of organs and tissues, they

Table 1. Probable death rates in USA.

Cancer	Estimated cases		Estimated deaths	
	Male	Female	Male	Female
Pancreas	23,530	22,890	20,170	19,420
Stomach	13,730	8,490	6,720	4,270
Liver	24,600	8,590	15,870	7,130
Lung	116,000	108,210	86,930	72,330
Breast	2,360	232,670	430	40,000
Thyroid	15,190	47,790	830	1,060
Eye	1,440	1,290	130	180
Intestine	4,880	4,280	640	570
Brain	12,820	10,560	8,090	6,230
Bone	1,680	1,340	830	630

Table 2. Estimated cancer cases and deaths across globe.

Cancer type	Estimated cases	Estimated deaths
Prostate	233,000	29,480
Breast	232,670	40,000
Lung	224,210	159,260
Colon	136,830	50,310
Melanoma	76,100	9,710
Bladder	74,690	15,580
Lymphoma	70,800	18,990
Kidney	63,920	13,860
Thyroid	62,980	1,890
Endometrial	52,630	8,590
Leukemia	52,380	24,090

are considerably more accurate in the imaging method for predicting stages of bone cancer [2].

Further, monitoring, treatment and assessment were done after radiation therapy (RT) and it was based on the output of CT and MRI [3, 4]. The noise from the images is not eliminated fully using this traditional bone feature extraction technique. Hence there is a need for automated and reliable techniques to carry out image analysis and these automated images produce digitized images of bone cross sections and further it leads to extraction of microstructure information [5].

Introduction to Image Processing

Image processing has been used in many scientific fields, such as in medicine or biology, however the researchers represent different types of cells by their texture properties [6], or distinguish between the alive or dead cells by analysing their images [7]. Edge detection is an important operation in image processing, which reduces the number of pixels and saves the image structure by determining the boundaries of objects in the it. The two general approaches to edge detection that are commonly used are gradient and Laplacian. The gradient method uses the first derivative of the image, and the Laplacian method uses the second derivative of the image to find edges. Pre-processing of an image is performed for the improvement of the image data and also for identifying image features which are important for further processing.

Introduction to Artificial Intelligent (AI)

Different AI algorithms have been used to carry out feature selection using a genetic algorithm. (GA) was well considered as a suitable evolutionary strategy for a large number of features and also applied to different areas, from object detection to gene selection in the example given with microarrays. GA, in general is used to estimate expression genes profiles with high predictive capability and biological relevance. Several algorithms have been developed for bone fracture detection [9]. It is important to estimate the amount of noise from the noisy image, then replacing the centre pixel by the mean of the sum of the surrounding pixels based on a threshold value. Clustering based on k-means is closely related to a number of other clustering and location problems. These include the Euclidean k-medians (or the multisource Weber problem) [10, 11], in which the objective is to minimize the sum of distances to the nearest centre and the geometric k-centre problem [11–13] in which the objective is to minimize the maximum distance from every point to its closest centre.

Aim and Objective

The main pinnacle is to predict the tumour intensity. In order to help the physician to diagnose and treat cancer, a fusion of techniques was involved in the detection process. Thus, hundreds of images were taken from the patient and the image segmentation procedure simplified the presentation and canalization. Further, disintegration of these wavelet based images and a rough guess for bone pictures were used in the recognition of cancer or malignancy, which enhances the visual nature of the bone through image processing and identifying the bone tumour. Finally a combination method was performed by merging the threshold segmentation and edge detection in order to get the precise segmentation. However, the data set of image processing was supported by the genetic algorithm.

2. Literature Survey

Cancer is the most spiteful disease, because prediction or detection plays a major role and also detecting its correct stage is more important for the therapists to treat the patient in an efficient manner. However, several techniques available at present review the detection of tumours.

Survey on Existing Method in Bone Cancer Diagnosis

The major cause for cancer deaths throughout the world is the forecasting level before the metastases stage. It has been estimated that 1.04 million new cases of lung cancer were diagnosed during 1990 [14] [Shetty 2005]. A bone scan demonstrates the possible evidence of bone metastasis before the lesions became evident [15] (Kim et al. 1984).

In the traditional methods, monitoring, treatment and assessment were done after radiation along with a CT or MRI scan [16, 17] (Rajer et al. 2008, Nestle et al. 2009).

Numerous algorithms were developed for detection of bone tumours [9]. Vijaykumar et al. in 2010 presented a filtering algorithm for Gaussian noise removal. Compared to other filtering algorithms, this gives a lower Mean Absolute Error (MAE) and higher Peak Signal to Noise Ratio (PSNR). Close estimation by wavelets is another instrument in science, physical science, and building algorithms [18].

Generally the DICOM images are corrupted by the salt and pepper noise [19]. Poisson and a Gaussian approach is used along with the inference of the mean and variance [20]. Chan et al. 1999 proposed curvelets transform. The Haar method gives the maximum accuracy value compared with the other two methods.

Wavelet thresholding de-noising system takes into account discrete wavelet transforms (DWT) proposed by Donoho and Johnstone, which are utilized by the denoising part of the ECG signal [21, 22] (Donoho et al. 1994, Michel Misiti et al. 2007).

3. Methodology

In our planned system, the input image was taken as a noisy image. Wavelet transform is applied to obtain level-based segmentation. Finally, the transformed image is converted into a noise free image by utilizing visu shrink threshold. The block diagram Fig. 1, represents the denoising by using wavelet transforms.

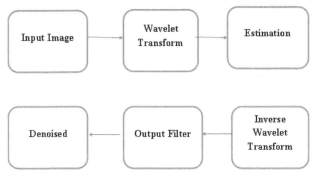

Fig. 1. Represents the denoising by using wavelet transform.

Two Dimensional Discrete Wavelet Transform

From the output of the wavelet transform method, we understood that an exceptionally inadequate picture was represented in tumour prediction. In this process, the wavelet algorithm was a useful contraption for signal processing and it also exhibited an adept representation of images. The image processing methods have created a remarkable interest on the grounds to permit expansive scale factual assessment not withstanding traditional eye screening assessment and are utilized as a part of both areas of pathology: cytology and histology.

Two dimensional (2D) Discrete Wavelet Transform (2D DWT) [23, 24] (Ta et al. 2009, Mauro Barni et al. 2006) is utilized as a part of image examination, denoising and division. The section channels in 2D DWT are utilized for further handling of the picture coming about because of the first step. This image decomposition [24] (Mauro Barni et al. 2006), process is as shown in Fig. 2.

The inspiration for utilizing the discrete wavelet is to split images into various portions based on variability. However due to low computing necessities, two dimensional signal processing is used. The original image was processed with wavelet transform analysis, shown in the Fig. 3.

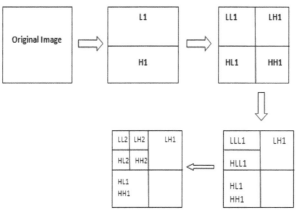

Fig. 2. Decomposition process.

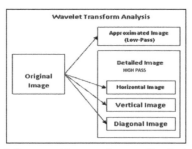

Fig. 3. Decomposition up to the second level scheme.

Effect of Threshold in Bone Cancer Diagnosis

The MRI Scan Image taken was from the Apollo hospital database, as it is important for diagnosing the tumour and for future treatment. In order to get a perfect image pre-processing activity is essential. As the algorithms were used to remove the noise, it is a difficult task to finally produce a perfect image for prediction of tumours and treatment. The previous methods used significantly removed the noise and produced a blurred image. To get a perfect denoising image we have introduced the wavelet algorithm, for image compression and denoising. We had performed wavelet threshold for the signal estimation strategy which was ultimately used for denoising.

By utilizing the hard threshold, coefficients that are slightly lower than the limit vanish and others are kept unaltered as shown in Fig. 4. The soft threshold made a significant mark, however immaculate noise passes the hard threshold, Fig. 5 and shows the irritating blips in the final yield which might lead to misidentification by the therapists as shown.

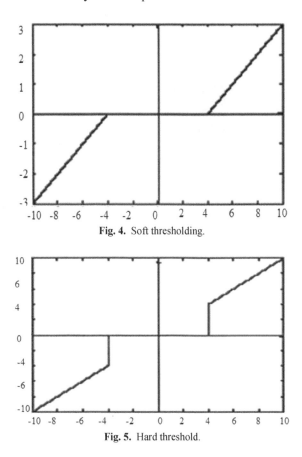

Fig. 4. Soft thresholding.

Fig. 5. Hard threshold.

The original image, threshold selection by VISU shrink is shown in Fig. 6. The denoised image at level 5 obtained by wavelet coefficients thresholding using fixed HARD threshold is shown in Fig. 7 and the histogram of original and denoised image performed by HARD threshold and unscaled white noise structure is represented in Fig. 8. Similarly, SOFT threshold was performed to get a denoised image using Haar at level 5 obtained wavelet coefficients and non-white noise structure are represented in Fig. 9. A histogram of the soft threshold applied to the MRI bone images for levels 1–5 are shown in Fig. 10.

Figure 7 shows the denoised image using Haar at level 5 obtained by wavelet coefficient threshold using fixed HARD threshold and scale white noise structure.

The Visu Shrink, represented by Donoho et al. 1994 [21] and Visu shrink threshold are used for estimating threshold. It utilizes a threshold value t that is corresponding to the standard deviation of the noise. However Visu

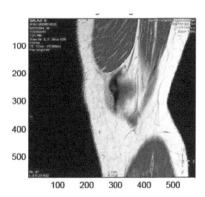

Fig. 6. Original image sample-1.

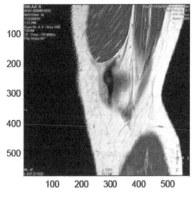

Fig. 7. Original image sample-2.

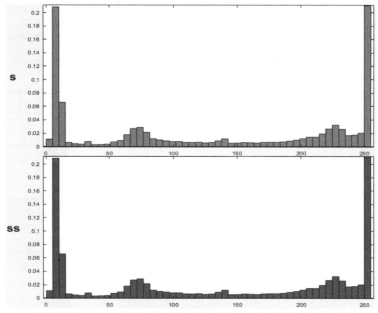

Fig. 8. Histogram of original and Denoising image using HARD threshold.

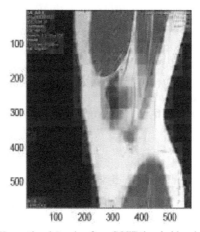

Fig. 9. Denoising image Haar at level 5, using from SOFT threshold and non-white noise structure.

Shrink does not contract with minimizing the mean squared noise. Another disadvantage, is, it cannot remove speckle noise, however, it can be used only with an additive noise. Therefore, we proposed that the input image be taken as a noisy image and transformed into various levels to convert to a denoising image as shown in Figs. 8, 9 and 10 respectively. From our finding, thresholds chosen have general limitations for hard and soft thresholding.

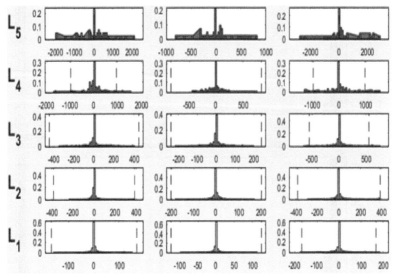

Fig. 10. Histogram of original and Denoising image using SOFT threshold.

4. Results

Wavelet Transform for Bone Cancer Diagnosis

From our methodology on the MRI scan bone image, wavelets based images were used to analyse the multiscale level in order to correlate the MRI output picture. The estimated level of 50 coefficients used for rough approximation, horizontal, vertical and diagonal input image were represented in Fig. 11 However, the modified decomposition of the original image was represented with Fig. 12. The accompanying preview demonstrates that the bone picture was fragmented at level 1 with close estimation coefficients. The related histogram and cumulative histogram marked the rough coefficient and we had tried for level 1 to level 5 as shown in Fig. 13. The approximation coefficient of bone images with level 2 to level 5 and their reconstruction with levels 1, 4 and 5 are represented in Fig. 14. From our findings, the rough guess gives a superior execution than points of interest of high recurrence. The Figs. 15, 16 and 17 depict the relating histogram and total histogram of bone images for levels 1, 4 and 5 respectively.

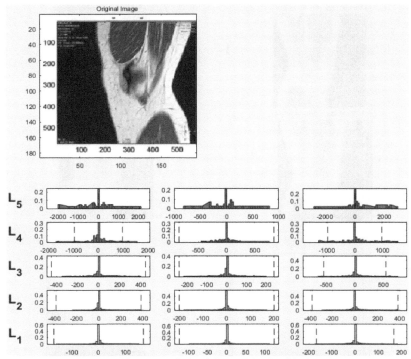

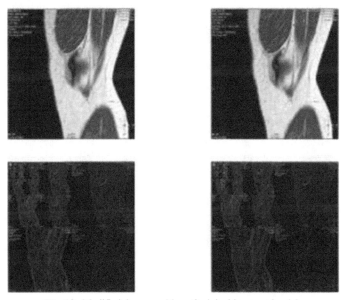

Fig. 11. Horizontal, vertical and diagonal details of the input image.

Fig. 12. Modified decomposition of original image at level 5.

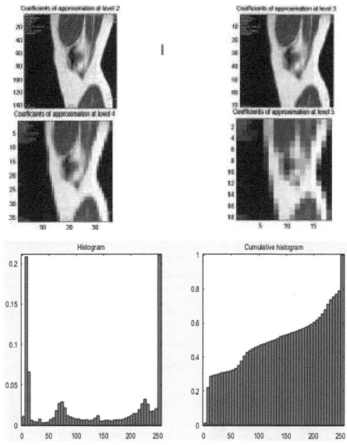

Fig. 13. Approximation coefficient of bone image at level 2–5.

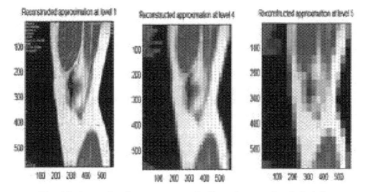

Fig. 14. Approximation reconstructed of bone image at levels 1, 4, 5.

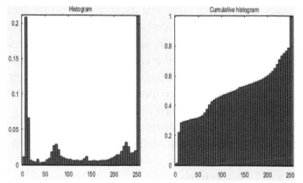

Fig. 15. Histogram and cumulative histogram of reconstructed approximation at level 1.

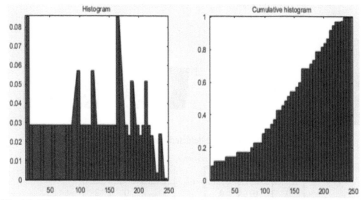

Fig. 16. Histogram and cumulative histogram of reconstructed approximation at level 4.

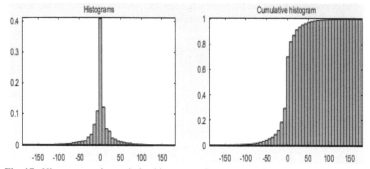

Fig. 17. Histogram and cumulative histogram of reconstructed approximation at level 5.

Results to Detect Bone Cancer Using K-Means Clustering and Edge Detection Algorithms

A novel approach, k means algorithm was performed for detecting the presence of bone cancer and subsequently to determine its stage with the computations of mean intensity and tumour size. The exploratory results demonstrate, the proposed system could get the smooth picture with edges representative of the ailment influenced part without spatial and spectral noises. K-means clustering segmentation also played a vital role. Subsequently, in order to get a vital property of the concentrate from the image, edge detection is essential.

The edge detector, proceeds by connecting high frequency points. Usually, an edge in the image will be a sharp intensity change intensity to detect the cancerous portion from the non-cancerous one. However, this is done using Sobel operator [25] (Om Pavithra Bonam and Sridhar Godavarthy). The results for boundary detection demarcated the diseased part from non-diseased part as in Figs. 18 and 19 respectively.

The output for K-means clustering gives the area of the tumor as in Figs. 20 and 21 respectively.

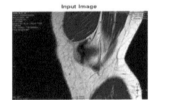

Fig. 18. Edge detected images of bone.

−1	−2	−1
0	0	0
1	2	1

−1	0	1
−2	0	2
−1	0	1

Fig. 19. The convolution kernel for the Sobel Edge detector. Note the emphasis on the horizontal and vertical edges.

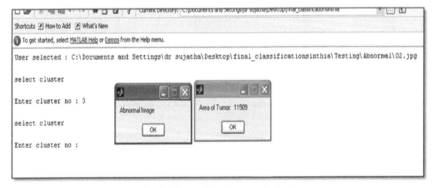

Fig. 20. Output for K-means clustering.

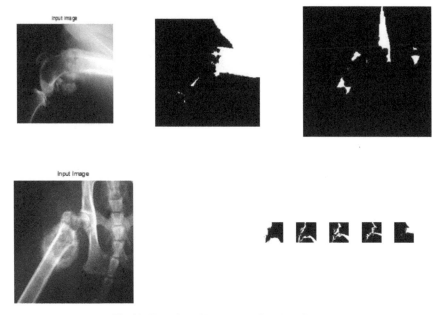

Fig. 21. Detection of Bone cancer from MRI images.

A Novel Approach to Detect Bone Cancer Using K-Means Clustering Algorithm and Edge Detection

With a continuation of the wavelet transform and co-efficient part in the proposed work, another approach has been incorporated with some pre-processing techniques. These techniques will be able to remove the noise to get a smooth image and will also increase the quality of the image, which is more suitable for segmentation as well as morphological operations and eliminates the false segments. Since we proposed the K-means algorithm to determine the stage based computations of mean intensity along with tumour size, edge detection was performed to get images without any spatial and spectral noise. The detection of Bone cancer from MRI images takes away the images that do not have a tumour or an unrelated image requires two main steps as in the Fig. 22.

Proposed System of Integrated Approach of K-means and Edge Detection Methods

The input noisy image was taken and transformed into various levels with the help of wavelet transforms. Then boundaries of objects were predicted with the help of the edge detection algorithm and further it was used for image segmentation and data extraction in the areas of image processing, computer vision and machine vision, as shown in Fig. 23.

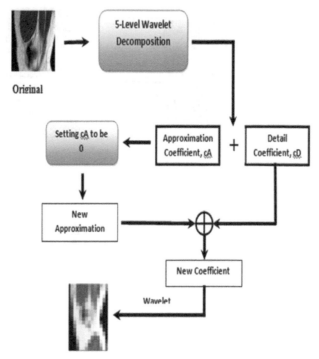

Fig. 22. Proposed schematic representation of decomposition process.

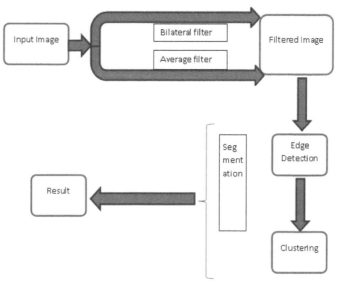

Fig. 23. Block diagram represents the integrated approach of K-means and edge detection methods.

Intelligent Classification Techniques

Genetic Algorithm

In the process of detecting the tumour, segmentation of medical images is a challenging part due to its poor image contrast and artifacts. Thus, it results in missing or diffusing the other tissue boundaries. Since these wavelet based GAs are proposed to detect the MRI images of the bone structure. The genetic algorithm (GA) is an exploration of the experimental part which mimics the development of natural evolution [28] (Banzhaf et al. 1998). As segmentation, can be taken as a method which catches out the optimal partition sections in an image according to a criterion, GAs are well altered to accomplish this goal as exhibited in Fig. 24.

However, the basic structure of a GA is shown in Fig. 25. As a start with an initial population, select parents from this population for mating. Apply crossover and mutation operators on the parents to generate new off springs. Finally, these off springs replace the existing individuals in the population and the process repeats. In this way a genetic algorithm actually tries to mimic the human evolution to some extent.

In the GA as show in Fig. 26, even after the process is ready there are other parameter to focus on such as population size, mutation and crossover probability in order to find the ones which suit the particular problem.

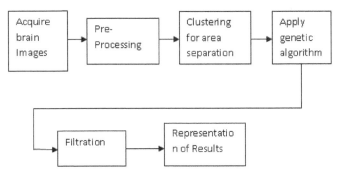

Fig. 24. Basic chart of tumour detection.

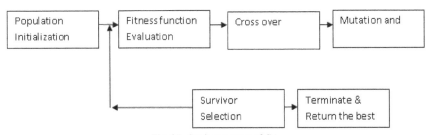

Fig. 25. Basic structure of GA.

From the output of GA, shown in the Fig. 27 among the 122 total selected samples, 42 showed benign, 40 showed malignant and 40 of them showed acute malignancy. The prediction of affected samples was accurate using the GA method. It was supported by the [29] Lipo wang, Feng chu et al. cancer prediction using gene expression data and it found minimum gene probability.

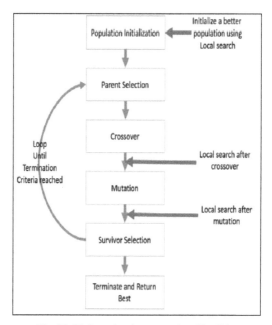

Fig. 26. Various cirertia were analysed by GA.

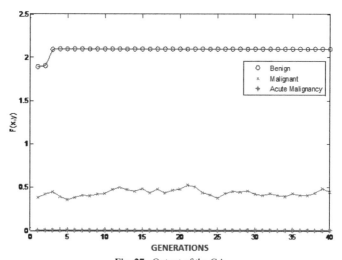

Fig. 27. Output of the GA.

5. Conclusion and Future Scope

A Bone tumour is an aplastic growth of tissue in bone and it was classified as a primary tumour which originates in the bone or from bone derived cells and tissues, secondary tumours originate in other sites and spread metastasis to the skeleton. With these known factors, the prediction and identification of the tumour stage is more important for therapists to provide effective treatment. In order to detect and identify the bone tumour, the normalized image was performed to get a defined standard image and enhancements over the image were done in the pre-processing stage. K-means clustering was performed to identify image areas and to predict the maximum chances of tumour. In the final stage, the genetic is implemented on this population set. With this specification, the genetics are initiated and processed by the algorithm with its continuous stages of selection, crossover, mutation, and more. After completing the genetics process, the valid threshold values were used to detect the tumour area in the bone image.

For analysis, a sample database of MRI scan images was diagnosed by using image pre-processing activity as the primary step followed by denoising. However, during the removal of noise, data should not be lost which is ultimately needed for detecting the tumour. It was found that a hard threshold was not always giving recovering denoising outputs since it depends on the wavelet thresholding algorithm. From our analysis, hard and soft on noisy versions of the standard 1D signals the threshold was discovered by using the Visu shrink. However, then the addition surface pictures need to be denoised to get an enhanced image as an output. In order to increase the quality, image segmentation was performed along with the K-means algorithm. The images without tumours or other unrelated images were removed by using average and bilateral filters. Further, a combination of threshold segmentation and edge detection were used to get a precise segmentation. However, the margin area was also removed without disturbing the object concentrate part. Finally, the image was taken up for the detection of bone tumours, however the tumour stage needs to be known for appropriate treatment. Among the 122 samples of diseased patients, using the GA we were able to predict the type of tumour. Thus using the proposed technology, the bone cancer was detected at its early stage with the appropriate area of the tumour.

References

[1] Zhi-Qiang Liu, Hui Lee Liew, John G. Clement, C. David L. Thomas et al. May 1999. Bone image segmentation. IEEE Transactions on Bio-Medical Engineering 46(5): 565–573.

[2] Heather A. Jacene, Sibyll Goetze, Heena Patel, Richard L. et al. 2008. Advantages of hybrid SPECT/CT vs SPECT alone. The Open Medical Imaging Journal 2: 67–79.

[3] Rajer, M. and Kovač, V. 2008. Malignant spinal cord compression. Radiol. Oncol. 42: 23–31.

[4] Nestle, U., Weber, W., Hentschel, M. and Grosu, A. L. 2009. Biological imaging in radiation therapy: role of positron emission tomography. Phys. Med. Biol. 54: R1–25.

[5] Detection of Bone Fracture using Image Processing Methods International Journal of Computer Applications (0975–8887) National Conference on Power Systems & Industrial Automation (NCPSIA 2015).

[6] Anand Jatti. 2010. Segmentation of microscopic bone images. International Journal of Electronics Engineering 2(1): 11–15.

[7] Amit Vasanji, PhD and Brett A. Hoover, MS. 2011. Image Analysis Techniques for Quantifying Bone In-Growth, Published: November 18, 2011.

[8] Vijaykumar, V., Vanathi, P. and Kanagasabapathy, P. 2010. Fast and efficient algorithm to remove Gaussian noise in digital images. IAENG International Journal of Computer Science 37(1).

[9] Arora, S., Raghavan, P. and Rao, S. 1998. ªApproximation schemes for euclidean k-median and related problems. º Proc. 30th Ann. ACM Symp. Theory of Computing, pp. 106–113, May 1998.

[10] Agarwal, P. K. and Procopiuc, C. M. 1998. ªExact and approximation algorithms for clustering.º Proc. Ninth Ann. ACM-SIAM Symp. Discrete Algorithms, pp. 658–667, Jan. 1998.

[11] Garey, M. R. and Johnson, D. S. 1979. Computers and Intractability: A Guide to the Theory of NP-Completeness. New York: W.H. Freeman.

[12] Fraser, A. 1957. Simulation of genetic systems by automatic digital computers. I. Introduction. Aust. J. Biol. Sci. 10: 484–491.

[13] Matousek, J. 2000. ªOn approximate geometric k-clustering.º Discrete and Computational Geometry 24: 61–84.

[14] Shetty, C., Lakhkar, B., Gangadhar, V. and Ramachandran, N. 2005. Changing pattern of bronchogenic carcinoma: a statistical variation or a reality? Indian J. Radiol. Imaging 15: 233–8.

[15] Kim, K., Kim, K. R., Sohn, H. Y., Lee, U. Y., Kim, S. K. et al. 1984. The role of whole body bone scan in bronchogenic carcinoma. Yonsei Med. J. 25: 11–7.

[16] Rajer, M. and Kovač, V. 2008. Malignant spinal cord compression. Radiol. Oncol. 42: 23–31.

[17] Nestle, U., Weber, W., Hentschel, M. and Grosu, A. L. 2009. Biological imaging in radiation therapy: role of positron emission tomography. Phys. Med. Biol. 54: R1–25.

[18] Srividhya, V., Sujatha, K., Ponmagal, R. S., Durgadevi, G. and Madheshwaran, L. 2020. Vision based detection and categorization of skin lesions using deep learning neural networks. Procedia Computer Science [International] 171: 1726–1735.

[19] Sujatha, K., Jayalakshmi, S., Sinthia, P., Malathi, M., Ramkumar, K.S., Cao, S.-Q. and Harikrishnan, K. 2018, Screening and identify the bone cancer/tumor using image processing. Proceedings Of The 2018 International Conference On Current Trends Towards Converging Technologies, ICCTCT 2018.

[20] Bhavani, N. P. G., Sujatha, K., Karthikeyan, V., Shoba Rani, R. and Meena Sutha, S. 2020. Smart imaging system for tumor detection in larynx using radial basis function networks. Advances in Intelligent Systems and Computing [International] 1108: 1371–1377.

[21] Donoho, D. L. and Johnstone, I. M. 1994. Ideal spatial adaptation via wavelet shrinkage. Biometrika 81: 425–455.

[22] Michel Misiti, Yves Misiti, Georges Oppenheim and Jean-Michel Poggi. 2007. Wavelets and their Applications. Published by ISTE 2007 UK.

[23] Ta, V.T., Lezoray, O., El Moataz, A. and Schupp, S. 2009. Graph-based tools for microscopic cellular image segmentation. Pattern Recognition 42(6): 1113–1125.

[24] Mauro Barni. 2006. Document and Image Compression. CRC Press, Taylor and Francis Group.

[25] Om Pavithra Bonam and Sridhar Godavarthy. "edge detection" Computer Vision (CAP 6415 : Project 2).

[26] Jay Patel and Kaushal Doshi. 2014. A study of segmentation methods for detection of tumor in brain MRI. Advance in Electronic and Electric Engineering 4(3): 279–284. 9.

[27] Jing, M. A. and Qiang, B. I. 2012. Processing practise of remote sensing image based on spatial modeler. The National Natural Science Foundation of China, IEEE 2012.

[28] Banzhaf, Wolfgang, Nordin, Peter, Keller et al. 1998. Genetic Programming: An Introduction, Morgan Kaufmann, an Francisco, CA.

[29] Lipo wang, Feng chu and Wei Xie. 2007. Accurate cancer classification using expressions of very few genes. IEEE/ACM Transactions on Computational Biology and Bioinformatics 4: 40–52.

Chapter 7
Swarm Optimization and Machine Learning to Improve the Detection of Brain Tumor

Anjum Nazir Qureshi

1. Introduction

Good health is the key to a happy and prosperous life. Healthy people are more productive and live longer. Every person on this globe wants to stay fit and healthy. A healthy person is able to accomplish all his tasks, serve his family, friends and the nation. Balanced diet, exercise routine and good hygienic habits are essential to maintain a healthy life. Along with this a regular routine check up with the family doctor or tests at the nearest clinic are required to monitor one's health. But when a person suffers from an illness he becomes dependent on others. The disease makes him weak and he is unable to fulfill assigned responsibilities assigned. It is not only the sick person who gets affected but also the people and society around.

The developments in the health care systems in the last few decades have revolutionized the availability of medicines, diagnostic facilities and treatments based on the test results. It has therefore increased the overall life expectancy worldwide and at the same time has increased the chances for health problems that occur with age. Moreover, the changes in lifestyle, eating habits and behavioral patterns have increased the risks of chronic ailments like cardiovascular diseases, diabetes, hypertension and cancer that require long term care. Globalization and urbanization are some of the factors that have given rise to risk factors like smoking, alcohol consumption, fast food and sedentary lifestyle that has contributed to the increase in the range of

Assistant Professor, Rajiv Gandhi College of Engineering Research & Technology, Chandrapur.

chronic diseases and has affected every age group. A number of organizations are working to disseminate awareness to reduce the exposure to poor diet and sedentary lifestyle to save lives and reduce the health care expenditure of the world. Increased screening, regular checkups and treatment at the early stages can be effective in helping a patient to lead a normal life.

Cancer is one of the chronic diseases that has disturbed millions of lives worldwide, and, considered to be one of the largest health problems and the second leading cause of death. Cancer can affect any part of the body. The cancer cells group together to form tumors which destroy normal cells around them and damage the healthy tissues of the body and make a person very sick. Cancer can be cured if detected at an early stage and following the medical treatments given by the doctors. Early diagnosis thus plays a crucial role in the treatment of cancers. The advancements in technology have been helpful in providing some fast and reliable methods to diagnose cancers and enhance the obtained results so that the treatment can be started without delay. This chapter will discuss about some the latest technologies that are helpful in improving the speed and accuracy of results obtained by using Magnetic Resonance Imaging (MRI) for brain tumors, which is a type of cancer.

A collection of abnormal cells within the brain is called a brain tumor. The tumors can be cancerous or non-cancerous. A neurologist or a neurosurgeon may use computer tomography (CT) scan or magnetic resonance imaging (MRI) for diagnosis of a brain tumor. The test results are then used by the neurologists to prescribe treatment for their patients. Segmentation of medical images for detecting the exact size or location of the tumor is challenging due to the intrinsic nature of images. There are a number of segmentation methods that have been developed to overcome the difficulties faced by radiologists and doctors to detect the abnormalities in the medical images due to noise or due to spatial variations in illumination. Particle swarm optimization (PSO) techniques are being used by researchers to lessen the affects of challenges faced while detecting brain tumors from MRI images and reduce the time required to give exact results of the tests. PSO when combined with other soft computing techniques like Fuzzy and genetic algorithms can improve the quality of segmentation for the MRI images. Though beneficial for the people who need them but it may require a long time, approximately one hour to complete the whole process. Machine Learning (ML) can reduce the time needed for producing MRI images. The ML methods can be used for analyzing the available historical data of the tumors and comparing it with the current MRI image. This would help the doctors in obtaining better insights about the patient's tumor; feel more confident to make the diagnosis and to reduce potential mistakes to avoid improper treatment plans. This chapter will discuss some of the Swarm optimization techniques that are used to improve the results of MRIs for brain tumors. It will also discuss the ML algorithms used for brain tumor detection and its benefits in the diagnosis process as compared to the traditional methods.

2. Brain Tumor and its Diagnosis

All living creatures are composed of cells. New cells are formed by the human body to replace the old and damaged cells. Formation of new cells is more in infants and children to complete their fast growth process. A tumor starts developing in a body when the normal or the abnormal cells multiply at a higher pace than required for the replacement of damaged cells. An unnecessary mass of cells starts developing in the brain is termed as a brain tumor.

2.1 Types of Brain Tumors

The handbook by American Brain Tumor Association [1] classifies the brain tumors basically into two types: Primary and Metastatic and assigns grades to the tumor ranging from 1–4 depending upon the treatment difficulty levels.

The primary brain tumor is the one that starts developing in the brain. Though there are more than 120 types of primary brain tumors that exist, these tumors are broadly classified as benign and malignant. The various kinds of primary tumors are assigned names according to the kind of cells involved.

The benign tumors exhibit a normal cellular appearance when viewed under the microscope. The borders do not spread, are visible distinctly and grow very slowly. Surgery is considered to be the most effective treatment for benign tumors. The benign tumors can be life threatening if they are located in vital parts of the brain as the surgery process may impact the way of working

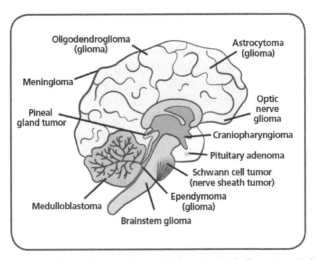

Fig. 1. Primary brain tumor (source: Handbook, American Brain Tumor Association [1]).

of the vital areas. On the other hand the brain tumors that grow rapidly are called malignant. The malignant tumors are life threatening, sometimes also referred to as brain cancer. These tumors rarely spread to other parts of the body but can spread within the brain and spine. As these tumors send roots to the normal adjoining tissues no distinct borders can be found in them. They shed cells that reach the distant parts of brain and spine through the cerebrospinal fluid.

The secondary or the metastatic brain tumors start in some other body parts and then spreads to the brain. This type of tumor is more common among adults as compared to adults and found mostly in people with a history of cancer. Some of the cancers that spread to the brain include breast cancer, colon cancer, kidney and lung cancer.

The medical professionals from the World Health Organization (WHO) have assigned grades to the tumor to ensure better communication within the healthcare team and to facilitate planning for treatment of the patients and to anticipate its results. The grades are assigned as 1–4 depending upon the severity of the tumor. The grade-1 tumor is considered to be the easiest for treatment while grade-4 to be the toughest. The grade-1 tumors grow

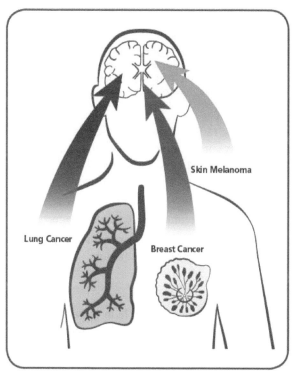

Fig. 2. Metastatic tumor (source: Handbook, American Brain Tumor Association [1]).

gradually, are considered to be the least cancerous and correlated with long term survival. The microscopic view of these tumors looks like normal cells and can be treated with surgeries. The grade-2 tumors appear slightly abnormal in the microscopic view and grow slowly. They can affect the adjoining normal tissues, reproduce themselves and develop into a higher grade tumor. The cells of grade-3 tumors appear to be abnormal in the microscopic view. These cells divide rapidly and are likely to spread in the nearby tissues. The grade-4 tumor appears to be very abnormal, divides actively and spreads very quickly. It is considered to be the most difficult type to be treated. It forms new blood vessels for fast growth of the tissues and has areas of dead tissues in the centre.

2.2 Causes and Symptoms

The symptoms for brain tumors may vary from person to person. It also depends on the size and location of the tumor. Some of the common symptoms as enlisted in [3] are vomiting, blurred vision, confusion, memory loss, weakness of limbs, difficulty in walking, mood and behavioral changes, loss of balance and loss of bladder control. The patients may commonly get headaches which may get worse by coughing or sneezing, exercise and may be worse while waking up in the morning.

The exact causes of brain tumors are still unknown despite extensive research. Some of the causes and risk factors discussed in [3] are:

- Some of the brain tumors can be genetically inherited. The risk of developing a tumor increases if many family members have suffered from it in the past.
- Exposure to ionizing radiation increases the risk of tumors. The radiation may be due to X-rays or CT scans, nuclear plants, power lines and mobile phones. Similarly repeated exposure to harmful chemicals like nickel or cadmium compounds, tobacco smoke and arsenic compounds may put a person to the risk of developing a brain tumor.
- The risk of getting a brain tumor increases with age.
- Obesity or being overweight is also considered to be one of the risk factors in getting some types of brain tumors.

2.3 Diagnosis

Diagnostic steps play and an important role in disease management and help in determining a specific condition a patient is suffering from. When a doctor observes symptoms like long term headaches and loss of balance, he may prescribe scans for the brain and suggest the patient to visit a neurologist or neurosurgical oncologist for further treatment. A scan helps to determine the presence of a tumor, its location and size. Sometimes the patient may require more than one type of scan for better diagnosis of its type and location. Some of the commonly used scanning techniques are CT scans and MRIs [2, 4].

CT-scans use a series of X-ray images captured from different angles and computer processing for combining these pictures for creation of 3-dimensional images of the tumor. These images can be used to find bleeding and enlargement of the fluid filled spaces. Besides this seeing the changes in the bones of the skull and assessing size of a tumor can be done with the CT images. A CT scan is generally recommended to a patient who cannot go for a MRI like a person who has a pacemaker heart transplant. A dye or a contrast medium is injected into the patient's bloodstream before the scan to obtain better details with the image. Positron Emission Tomography is combined with a CT scan, generally known as PET-CT scan which uses a small amount of radioactive substance to create images of activity in the tumor cells. The tumor cells tend to divide faster as compared to the other tissues and therefore absorb more amount of the injected radioactive substance. The scanner detects this substance for creating images of the tissues.

A MRI uses magnetic fields to generate detailed images of a body part. A dye or contrast medium is used for getting a clear image of the scan. The dye may be injected into the patient's body or may be given as a pill or liquid. A MRI is preferred to a CT scan as it produces better images. It is also used to detect the size of the tumor. There are different types of MRI techniques which are recommended by the doctors according to its type and its possibility of spreading to other parts of body. One of the types is a Prefusion MRI that is used to check the flow of blood into the tissues. This process helps the doctors to predict the affect of treatment and to distinguish between a recurrent tumor and a dead tumor tissue. A Functional MRI (FMRI) is used by the doctors to plan for the surgery. During this scan a person is asked to perform certain tasks and the changes caused in the brain due to these tasks, like use of oxygen and blood flow in brain, are studied by the doctors. FRMI can therefore help the surgeon avoid damage to functional parts of the brain while removing a tumor. Magnetic Resonance Spectroscopy (MRS) is used to obtain information about the chemical composition of the brain or the number of metabolites in the body. This test helps to evaluate the response of therapies by differentiating among the dead tissues caused by previous treatments and new tumor cells.

In cases where the results of the scan are unable to provide appropriate results another procedure called biopsy [3] is used to determine the type of tumor. In this procedure a small number of tissues is removed which are examined by the pathologist under a microscope to diagnose the exact type of tumor. The tissues can be removed using a needle biopsy in which a hollow needle is guided into the tumor through a small hole that is drilled into the skull. Biopsy can be also used as a part of surgery for removing the tumor.

Similarly, a stereotactic biopsy uses a stereotactic head frame and computer to direct the needle to determine the exact location of the tumor. This procedure is used for tumors that are located deeper in the brain.

3. Swarm Intelligence Algorithms for Brain Tumor Detection from MRI Images

Swarm intelligence (SI) is an algorithm that is biologically inspired. It has been developed considering the behaviors of social insects like ants and bees. It is a branch of Artificial Intelligence (AI) that is used to model the collective behavior of natural swarms like bird flocks, colonies of ants and bees [6]. The principles of SI have been effectively used by researchers for numerous problem domains like the analysis of an image or data, structural optimization and optimal route selection [7]. Some of the examples of SI models that use the behavior of natural swarm systems for real life applications are Particle Swarm Optimization (PSO), Artificial Bee Colony, Ant Colony Optimization (ACO), Bacterial Foraging and Artificial Immune Systems [5]. Among the various SI algorithms, the ACO and PSO are widely used for the tumor detection from the MRI images. These SI algorithms when combined with Fuzzy and Genetic Algorithms (GAs) give improved results to facilitate the detection of a tumor, its location and size from its MRI images.

3.1 PSO Algorithms for Brain Tumor MRIs

PSO was introduced by Rusell Eberhart and James Kennedy in 1995 [8]. It was earlier used to solve non-linear continuous optimization problems but later the model was developed for solving practical or real-life problems. PSO is inspired by the characteristics of bird flocking in which the birds fly long distances in large groups by maintaining appropriate distance with their neighbors to avoid collision [5]. The basic PSO algorithm uses a velocity vector to update the current position of each particle is swarm. The update for the position of each particle is carried out on the basis of social behavior that is adopted by the population of individuals or the swarm. The memory of each particle and the knowledge acquired by the swarm is used for updating positions. The basic steps involved in the PSO algorithm are [10]:

  i. Begin with a set of particles that are randomly scattered all over design space.
 ii. Compute the velocity vector for each particle of the swarm.
iii. By using the previous position of the velocity vector and the updated vector, the position of each particle is updated.
 iv. Return to step 2 and repeat the process until convergence.

The different types of tumors with varying sizes and shapes make it difficult for the radiologists or the doctors to detect the tumors from the MRI. To deal with this the authors in [9] proposed a clustering algorithm based on PSO. The MRI images in the given model are classified on the assumption that different types of features exhibit distinct values for the pixel depending upon spectral reflectance and emittance properties. The classification based on the spectral information of each pixel or value of radiance that is attained at different wavelength bands for each pixel is called, spectral pattern recognition. In addition to this spectral and temporal pattern recognition are also applied to the MRI images. This study was conducted on a set of normal as well as abnormal MRI images obtained from hospitals. For all these images Gray level co-occurrence matrices are built by extracting the pixels and using the spatial coordinates and intensities. The algorithm locates the centroids for the clusters, where the clusters group together tumor patterns obtained from the MRI images.

The segmentation of the tumor from the MRI images is challenging due to low contrast, unclear boundaries and lack of accuracy. This problem can be solved using computerized or automatic methods for the segmentation and detection of tumors. The Enhanced Darwinian Particle Swarm Optimization (EDPSO) used in [11] uses an automated segmentation process that yields better results than the PSO. The algorithm is implemented in four steps. In the first step called, preprocessing, the unwanted artifacts like personal details of the patient (name, age, sex) and noise are removed using a tracking algorithm. In this step the image is stored in a two-dimensional matrix. A peak threshold value is set to remove the white labels. The flag value is set to 255 after which the pixels equal to 255 are selected and the rest are removed. In the second step, called, image enhancement, noise (salt & pepper noise, impulse noise) and high frequency components are removed using a Gaussian filter. After this the segmentation is done using EDPSO in the third step. The EDPSO particles travel through the search space for finding the best solution by interacting and sharing information with the neighboring particles. After every step the best solution called, the neighborhood best, is updated by which the particles know the location of the search space to obtain success and are lead by these successes. Classification is done in the fourth step by using an Adaptive Neuro Fuzzy Inference system that uses Artificial Neural Networks (ANNs) and fuzzy logic. This algorithm is easy to implement, provides better classification of the images and requires less time for execution.

Another Fuzzy based approach for segmentation of the brain tumor from the MRI images has been proposed in [13] that performs image segmentation of the brain MRI images using adaptive thresholding. The MR brain image is classified into two membership functions (MF) for calculation of fuzzy entropy. Gray scale images with 256 gray levels having membership functions corresponding to the dark and bright pixels are considered for the experiment. The threshold value depends on the parameters a, b, and c and their values lie in the range of 0–255 which are used to design fuzzy MF's. The two membership functions used are Z-function that denotes the class dark and S-function that denotes the class bright are constructed using the values of a, b and c. The PSO is initialized for the image matrix for which the fitness value is calculated for each particle using the fuzzy entropy function. After the calculation of the fitness value it is compared with the previous ones to determine the best current and previous position. If the current value is better, the velocity and the position of all particles are updated. After calculation of fitness values for all particles, the threshold value is determined based on the values of a, b and c. After this step segmentation of the image is done that displays only the tumor regions from the MRI images.

One more algorithm for segmentation of tumors from the MRI images by using PSO has been discussed in [12] that works in four stages. In the first stage the obtained MRI files are converted into image files that can be read with the imaging software on a laptop or computer. In the second stage PSO is applied on the image files. The segmentation starts with the segmentation level, n = 2. The values of segmentation levels are changed to 3, 4, 5, 6, 7 and 8 for the axial as well as the coronal plane respectively.

In the third stage the resultant image is derived by using the elapsed time of both the axial and coronal planes. The elapsed time values of all the segmentation levels are compared and the best value is selected as the resultant image that gives the accurate result. In the fourth stage the noise and artifacts are removed to display only the tumor regions from the resultant image as shown in Fig. 3(b) for the axial plane and in Fig. 4(b) for the coronal plane. Therefore, with the help of the algorithm the exact location and size of tumor could be known very quickly which would be difficult to do manually.

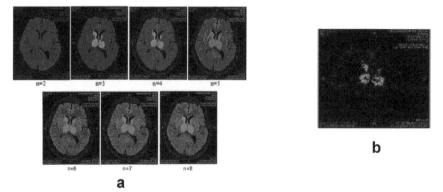

Fig. 3. (a) PSO algorithm in axial plane (b) Segmented Brain Tumor in axial plane [12].

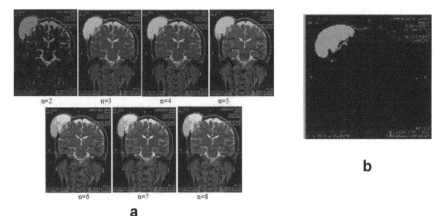

Fig. 4. (a) PSO algorithm in coronal plane (b) Segmented Brain Tumor in coronal plane [12].

A modified PSO (MPSO) technique in [14] detects the tumors precisely in less time for the 2D MRI images. The methodology used in this research comprises of five steps: acquisition, preprocessing, segmentation, reconstruction and performance testing. The image acquisition process consists of getting 3D MRI images. In the preprocessing stage, these acquired 3D images are converted into 2D image sequences. Multiple 2D slices can be obtained from the 3D images that can be used for the segmentation process. For the segmentation of images, the MPSO algorithm works in few steps. The step 1 includes initialization of particles with random positions and velocity vectors. In step 2 the fitness value is determined for each particle position. The obtained value is compared with the threshold or the best value that has been set by the algorithm in step 3. In the next step the value of fitness is updated and again compared with the threshold and the updating stops once the best value is obtained. The reconstruction stage helps to extract the tumor from the 2D slices and then concatenates the slices to convert it into the original 3D images. The final step is the performance testing where obtained results are checked for accuracy, sensitivity and specificity.

In this section some of the PSO algorithms were discussed. Many other variations of the PSO algorithms for the segmentation of MRI images have been proposed by researchers. All the algorithms that have been covered in this section and also those which couldn't be included clearly indicate the benefits of using PSO for the MRI images. The MRI images consist of noise, artifacts and the neighboring tissues. The segmentation results obtained by using PSO are more precise due to which the doctors can easily know about the location of tumor, its size and its location in the brain. It can thus save time required by the doctors in the diagnosis. These algorithms can thus be a boon as the treatment of the patients will not be delayed and lives of many people can be saved.

3.2 ACO Algorithms for Brain Tumor MRI

Ant Colony Optimization (ACO) is a swarm intelligence algorithm that has been developed using the foraging behavior of ants. The ants deposit pheromone on the ground while moving from one place to the other and rest of the ants follow the same path. This mechanism of the ants has been used by the ACO in solving optimization problems. It uses a number of artificial ants that determine solutions to the given optimized problems. The well known travelling salesman problem uses ACO to find the shortest path to the given set of cities [15]. ACO is based on iterations. For any ACO algorithm, a number of ants are considered for every iteration. The artificial ants find a solution by traversing the fully constructed graph. The vertices of the construction graph represent the solution components on which the pheromone is deposited. At each iteration, a number of solutions are constructed by the ants. Local search is then used to improve the obtained solutions. Applying local search is problem specific and optional but used in most of the ACO algorithms. After this the pheromone values are updated, by which the pheromones related to the promising or good solutions can be increased and those related to the bad solutions can be decreased. The good results of ACO algorithms have made them appealing for numerous applications [16]. This section will discuss, ACO algorithms used for the segmentation of MRI images.

3.2.1 ACO and Fuzzy

Brain tumor diagnosis using ACO and fuzzy has been discussed in [17, 19] in which the artificial ants construct a solution by utilizing the pheromone information accumulated by other ants. Hierarchical Self Organizing Map (HSOM) has been used for the segmentation process that helps in the selection of clusters from the given image. ACO and fuzzy are then used to extract the suspicious regions. An initial pheromone value T_0 is assigned for each ant and a random pixel is chosen from the image that has not been used previously. Every pixel is assigned a flag value 0 initially. Once it is used, its flag value is updated to 1. The flag value helps to learn if the pixel has been used earlier. This method is followed for all the ants. A separate column is allocated for updating the flag and pheromone values of each ant. A local pheromone is done to compute the posterior energy. In every iteration, the local minimum value is selected from the ant's solution and a local pheromone update is performed. A local minimum value is calculated using Fuzzy. If the value of the local minimum is less than the global minimum, it is assigned to the global minimum. A Pheromone update is done for the ant that generates a global minimum value. Thus, the global minimum value returns the desired level of image segmentation at the final iteration. This segmentation method enhances the image and gives better results that help the observer to perceive the region of interest (tumor in this case) better as compared to manual segmentation.

3.2.2 ACO and K-means

One more approach using ACO and K-means clustering for the diagnosis of brain tumor has been studied in [18, 19] that gives very clear and good pictures of the segmented brain images. Image segmentation is considered to be one of the toughest part of image analysis in which the areas having pixels with similar attributes are extracted. Similarly, segmentation of tumor images is challenging as compared to the natural images due to the high functional sensitivity of the MRI images. The pheromone distribution is used here for determining the threshold value that helps in improving the quality of segmentation of the MRI images. The impulsive noise in the image is removed by using median filter and the large size images are resized in the preprocessing stage. In the first part of the algorithm a pheromone matrix is constructed depending upon the tendency of the ant to move to different orientations. The probability of the movement of ants depends upon the pheromone intensities in other sites. An ant is assigned to each pixel. The ant moves towards the neighbor pixel for a certain number of steps. The pheromone value of the pixel is updated with every step of the ant. A binary image is obtained by normalizing the pheromone matrix and the primary image. K-means clustering method is then applied for segmentation that segments the image by forming a cluster of data points with similar criterion. The execution time is very less and the tumor region detected by the algorithm is distinct and clear without any extra margin that may sometimes be observed due to inflammation.

K-means clustering that had been used for segmentation in the earlier research works, allows clustering of pixels belonging to one class but it does not work well for the images that have clusters of various sizes and densities. To overcome this, the authors in [21] proposed Fuzzy c means (FCM) algorithm that allows pixels to belong to different clusters with different degrees of membership. In this algorithm the artificial ants are propagated uniformly over the MRI image space to perform the search activity. Gradually the entire image is covered by the searching ants and the pheromone update is saved in the form of a histogram curve. The entire image falls into three possible scenarios depending upon the group of image pixels: background, target and boundary of target. Two types of updates are performed over the pheromone matrix. The first kind of update called the local pheromone update is done by all the ants after each construction step till the last step is reached. The second type of update known as the offline pheromone update is done after each ant has completed one cycle of iteration. The probability of an ant to move in a particular direction depends on the amount of pheromone on a path.

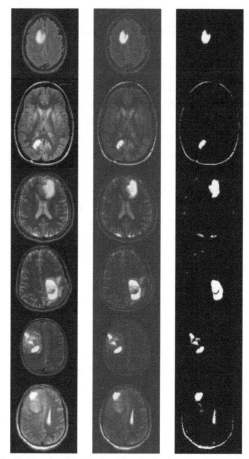

Fig. 5. Results of ACO algorithm for brain tumor images [18].

The pheromone evaporation enables the ants to explore a new and shorter path that prevents the system from following substandard paths that are mostly developed due to a large concentration of pheromone on a path. The computation time of this algorithm is less due to the determination of a shorter path and the quality of segmentation is better than the K-means methods.

3.2.3 Bee Colony Algorithm

The Artificial Bee Colony (ABC) algorithm was first proposed by D. Karaboga in 2005 for optimizing real world and numerical problems. It is inspired by the foraging behavior of the honey bee swarms [24]. The three basic components of the algorithm are employed bees, onlookers and scouts. The employed bees search for the food sources and share this information with the onlooker bees which then select the food sources of high quality while the scout bees do a random search for food. The position of the food source indicates possible solutions to the given problem. The positions of the food source (solutions) are continuously updated till the given conditions are satisfied. The employed bee remembers its previous best position and creates new positions in the memory within its neighborhood. On detection of a better new food source the old position is updated with the new one. Once all the employed bees finish the search process the information of the route of the food sources and its length is shared with the onlookers through a waggle dance. The onlooker bee selects the food source by observing the waggle dance on the basis of the probability value of the sources. It then explores the area within its neighborhood for generation of new candidate solutions. In case no improvement is observed in a position after a fixed number of cycles, the position is discarded and the corresponding employed bee converts into a scout. A fresh randomly generated food source replaces the discarded position [25].

A segmentation method based on ABC has been discussed in [28] for extracting tumors from the MRI brain images. The whole process goes through three phases: preprocessing, processing and post processing. In the preprocessing stage, noise is removed by using filters like average, median and Guassian Adaptive filter. In the processing stage the ABC algorithm is used for segmentation of the images. The image is converted into binary images by using gray level thresholding. The tumor region is then extracted from the segmented image by using connected component labeling on the basis of pixel connectivity. The results obtained by this algorithm are better as compared to other algorithms based on K-means and FCM as they fail to separate tumor regions from other parts.

ABC algorithm has been used in [26] for segmentation of brain tumor from the MRI images. A three level discrete wavelet transform (DWT) is used to decompose the original image into approximation and gradient images. The low frequency components carrying approximation information are reorganized to form an approximation image while the high frequency

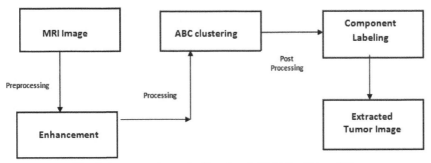

Fig. 6. A segmentation method based on Artificial Bee Colony [28].

components carrying edge information are reorganized to form a gradient image. The approximation image is filtered using a low pass filter. The gradient and the filtered image are normalized to [0, 255]. Improved two dimensional grey entropy, considered as the fitness function of ABC algorithm is obtained by using a 256 × 256 filtered gradient occurrence matrix. It is used to set the control parameters like population size and the number of iterations. An optimum threshold value is obtained after a number of cycles that is used to segment the filtered image. In the second stage FCM algorithm is used for clustering that enhances the tumor part in the segmented image and indicates its intensity.

Fuzzy clustering algorithms tend to get stuck in local minima and low convergence rates. To avoid this drawback of FCM an improved algorithm that combines ABC and FCM has been proposed in [27] that automatically segments the MRI brain images for determination of accurate location and appropriate number of tumor clusters and cells from the real as well as the normal brain MRI images. The proposed algorithm is able to overcome the drawbacks associated with the FCM algorithm and due to fast execution of the algorithm the results are obtained in shorter duration. The basic difference between the hybrid FCM ABC algorithm used here is that it uses a matrix to increase the speed up the calculation of the probability value. The algorithm segments the MRI images into number of clusters that use the intensity values of the pixels of the cluster as future space. A quantization index is used to evaluate the outcomes and performance of the algorithm by using the classification accuracy rate that can be calculated on the basis of similarity between the clustered image and the ground truth image.

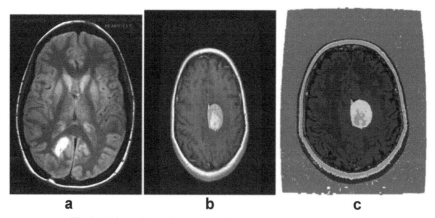

a   b   c

Fig. 7. (a) input image (b) segmented image (c) image after FCM [26].

4. Machine Learning for Segmentation of Brain Tumor from MRI Images

Machine Learning (ML) is a branch of artificial intelligence that is based on the concept that systems can learn, recognize patterns and do selections with minimum human interference. ML has gained momentum in the last few years due to the evolution of new soft computing techniques. It learns from previous computations for generating desirable and reliable outcomes [22]. Machine Learning is broadly classified as: supervised learning, unsupervised learning, reinforcement learning, deep learning and deep reinforcement learning [23]. In supervised learning the machine is trained using well labeled data. The machine is then provided with a new set of examples that use the supervised ML algorithms to analyze the labeled data and produce appropriate outputs from it. On the other hand, unsupervised learning algorithm train the machine to use unlabelled or unclassified data. It groups the unsorted data on the basis of some patterns or similarities. Reinforcement learning trains the ML models for making a sequence of decisions. In reinforcement learning the reinforcement agent learns from its experience and decides to perform a given task. The decisions are taken by the agent for maximizing rewards in a given situation. Reinforcement learning is generally used by the machines for detecting the finest solutions for the given problems. Deep learning is a type of machine learning that is inspired by the structure of a human brain. It uses a multi layer structure of algorithms by continuous analysis of data with a given logical structure. Deep reinforcement learning combines the deep and reinforcement learning. It can be extensively used to solve the tasks where complex decision making has to be done. Deep reinforcement learning is able to handle very large amounts of data and takes decisions for optimizing an objective.

4.1 Convolution Neural Networks (CNN)

A ML technique, convolution neural networks (CNN) has been used in [29] for the detection of a brain tumor, its classification and segmentation. The steps involved in the research are data collection, preprocessing, average filtering, segmentation, feature extraction and CNN via classification and identification. The image preprocessing stage improves the quality of the image by removal of noise and unnecessary data. Though the preprocessing stage removes noise from the image it is still not suitable for further processing. The average filtering stage provides an acceptable and smooth picture by replacing the pixel esteem with normal esteem. This stage thus removes salt and pepper disorder. CNN is used for image segmentation and for the extraction of the features from the pixel images. The CNN is an approach of deep learning that is used for image recognition applications. The two basic methods convolution and pooling are used to achieve high levels of classification. Classification involves the following steps:

i. Applying convolution filter.

ii. Minimizing the filter sensitivity by using sub sampling.

iii. The transfer of signal from one layer to the other is controlled by the activation layer.

iv. The rectified linear unit (RLU) is used to fasten the training period.

v. The neurons in the previous layers are associated with the neurons in the next layer.

vi. During the training the loss layer is used in the end to provide a feed back to neural network (NN).

The CNN uses parameters like root mean square error (RSME), recall, sensitivity and precision for generating results of classification of the images. The prediction and classification results from the MRI images are given as:

i. True Negative: Prediction for a patient that does not have brain tumor and was detected with no brain tumor.

ii. True Positive: Prediction for a patient having a brain tumor and detected with a brain tumor.

iii. False Negative: Prediction for a patient not having a brain tumor but detected for having a brain tumor.

iv. False Positive: Prediction for a patient having brain tumor but detected for having a brain tumor.

4.2 Support Vector Machines

A ML based approach has been discussed in [30] to help the physicians to study and diagnose tumors from the MRI images. The algorithm uses real time images for image classification and segmentation. The steps followed in the process are as follows:

i. The preprocessing step uses denoising algorithms in two stages. This technique uses wavelet domain filter for the removal of noise to obtain smoothness and better edge preservation. It uses linear filters like the median filter for the removal of salt and pepper noise while the non-linear filters like weiner are used for the removal of speckle noise. The images obtained after passing through filters is denoised using wavelet transform.

ii. In the second step the gray scale image in reduced to binary image. Ostu's binaraization method is used for performing cluster based image thresholding. K-means clustering is used for clustering the MRI image data. Feature extraction is done using wavelet transform.

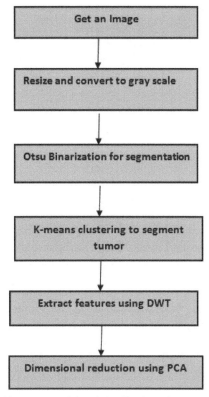

Fig. 8. Process of feature extraction and classification using K-means and SVM [30].

iii. Support Vector Machine (SVM) is used for the image classification. SVM has been preferred in this algorithm as it minimizes the bound on the errors made by the learning machine over the dataset used for testing. It also classifies the images not belonging to the training data. The classification is done to diagnose the tumor regions from the images obtained from the previous levels.

4.3 Supervised Classifiers

A ML approach for segmentation of brain tumors from the multimodality images has been proposed in [31] that uses wavelet based texture features for the extraction of tumors and supervised classifiers for classification. The process of segmentation is done in three steps: preprocessing, feature extraction and classification. The preprocessing stage determines the bounding box around the tumor region by removal of complete blank slices from the ground truth image so that only slices containing tumor parts are left and then creates a mask that is used to determine the bounding box in the ground truth.

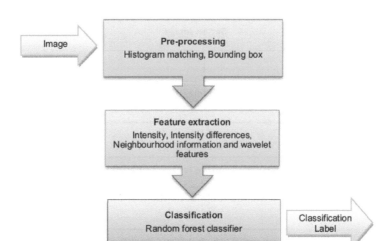

Fig. 9. Block diagram of the segmentation process for multimodal MRI images using wavelets and ML [31].

This bounding box is then utilized to crop the multi modality images. The feature extraction stage works on four types of features: Intensity, intensity difference, neighborhood information and wavelet based text features. The significance of wavelet based features is that it facilitates multi resolution analysis by which an image can be decomposed and visualized at different scales. In the next stage a ML based approach supervised classification is used for constructing the model and testing the data for evaluation of the constructed model on unseen data for testing the performance of algorithm. KNN (K-nearest neighbor), Random forest (RF) and AdaBoostM2 (adaptive boosting) are the classifier techniques used for the classification of images. At the end of the classification process the tumor is classified into three different regions: complete tumor, core tumor and enhancing tumor.

4.4 Deep Learning with Mixed Supervision

A deep learning based approach has been proposed in [31] to obtain better segmentation results than the standard supervised learning for the brain MRI images. The segmentation methods based on the ML models trained manually on segmented images require medical expertise and are time consuming. The deep learning model is trained for both fully annotated and weakly annotated data for performing as well as classification. The model is able to differentiate the three types of images: the first type that contains tumor with ground truth segmentation, the second type of images are the ones that do not contain a tumor and the third type that contain a tumor but do not have any provided segmentation.

The model is thus trained from training batches that include three types of images called K positive cases with segmentation, m negative cases and n positive cases but without any segmentation. The model is extended to the multiclass case in which image level labels are provided for each class. There is a possibility that an image can contain different types of subclasses. Therefore, this model considers one image level classification outcome for each tumor subclass so that the absence or presence of a given subclass is identified easily. Moreover, as each subclass has its own dedicated entire classification branch, the model is able to achieve better results. The mixed supervision model yields a significant improvement for the first two training scenarios.

5. Conclusion

Segmentation of the brain tumor from the MRI images is a challenging task and needs a lot of time for diagnosis. The swarm based techniques studied in this chapter are PSO, ACO and ABC while the ML based algorithms studied in this chapter are CNN, SVM, supervised classifiers. Deep learning techniques like swarm intelligence and ML are able to segment the MRI images efficiently by indicating very clearly the tumor, its size and location in the outputs. One more advantage of using these techniques is that the algorithms require less time for execution due to which the time required for manual diagnosis will be saved and the treatment of a patient can be started soon. Avoiding delays in the treatment can thus save a number of lives.

References

[1] Brain Tumors, A Handbook for The Newly Diagnosed, American Brain Tumor Association, Providing and Pursuing Answers, https://www.abta.org/wp-content/uploads/2018/03/newly-diagnosed.pdf.

[2] Brain Tumor, Frankly Speaking About Cancer, National Brain Tumor Society, Cancer Support Community, http://blog.braintumor.org/files/public-docs/frankly-speaking-about-cancer-brain-tumors.pdf.

[3] Brain Tumors: An Introduction, The IOWA Clinic, https://www.iowaclinic.com/webres/File/brain-tumors-intro.pdf.

[4] Allen Perkins and Gerald Liu. 2016. Primary brain tumors in adults: diagnosis and treatment. American Family Physician, Volume 93, Number 3, February 2016.

[5] Hazem Radwan Ahmed and Janice I. Glasgow. 2012. Swarm Intelligence: Concepts, Models and Applications, Conference: Queen's University, School of Computing Technical Reports, At: Kingston, Canada, Volume, Technical Report 2012-585.

[6] Yang, X.-S., Deb, S., Fong, S., He, X., Zhao, Y. et al. 2016. Swarm intelligence: today and tomorrow. 2016 3rd International Conference on Soft Computing & Machine Intelligence (ISCMI). doi:10.1109/iscmi.2016.34.

[7] Yan-fei Zhu and Xiong-min Tang. 2010. Overview of swarm intelligence. 2010 International Conference on Computer Application and System Modeling (ICCASM 2010). doi:10.1109/iccasm.2010.5623005.

[8] Kennedy, J. and Eberhart, R. C. 1995. Particle swarm optimization. pp. 1942–1948. In Proceedings of IEEE International Conference on Neural Networks, Perth, Australia.

[9] Chandra, S., Bhat, R. and Singh, H. 2009. A PSO based method for detection of brain tumors from MRI. 2009 World Congress on Nature & Biologically Inspired Computing (NaBIC). doi:10.1109/nabic.2009.5393455.

[10] Venter, G. and Sobieszczanski-Sobieski, J. 2003. Particle swarm optimization. AIAA Journal 41(8): 1583–1589. doi:10.2514/2.2111.

[11] Vasupradha Vijay, Dr. Kavitha, A. R. and Roselene Rebecca, S. 2016. Automated brain tumor segmentation and detection in MRI using enhanced darwinian particle swarm optimization (EDPSO). 2nd International Conference on Intelligent Computing, Communication & Convergence, Procedia Computer Science 92: 475–480.

[12] Mahalakshmi, S. and Velmurugan, T. 2015. Detection of brain tumor by particle swarm optimization using image segmentation. Indian Journal of Science and Technology 8(22): IPL0246, September 2015.

[13] Krishna Priya Remamany, Thangaraj Chelliah, Kesavadas Chandrasekaran and Kannan Subramanian. 2015. Brain tumor segmentation in MRI images using integrated modified PSO-Fuzzy approach. The International Arab Journal of Information Technology, Vol. 12, No. 6A.

[14] Gtifa, W., Hamdaoui, F. and Sakly, A. 2019. 3D brain tumor segmentation in MRI images based on a modified PSO technique. Int. J. Imaging Syst. Technol. 1–9. https://doi.org/10.1002/ima.22328.

[15] Marco Dorigo, Mauro Birattari and Thomas Stu"tzle. 2006. Ant colony optimization: artificial ants as a computational intelligence technique. IEEE Computational Intelligence Magazine | November 2006.

[16] Marco Dorigoa and Christian Blum. 2005. Ant colony optimization theory: a survey. Theoretical Computer Science 344: 243–278.

[17] Logeswari, T. and Karnan, M. 2010. An improved implementation of brain tumor detection using soft computing. 2010 Second International Conference on Communication Software and Networks, pp. 147–151. doi: 10.1109/ICCSN.2010.10.

[18] Soleimani, V. and Vincheh, F. H. 2013. Improving ant colony optimization for brain MRI image segmentation and brain tumor diagnosis. 2013 First Iranian Conference on Pattern Recognition and Image Analysis (PRIA), pp. 1–6, doi: 10.1109/PRIA.2013.6528454.

[19] Deepak A. Patil and Patil, S. N. 2016. Modified ant colony optimization algorithm for brain MRI image segmentation and brain tumor diagnosis. International Journal of Electrical and Electronics Engineers, Volume 8, Issue no. 02.

[20] Amalmary, M., Dr. Prakash, A. 2020. Fuzzy C-means clustering with ant colony optimization algorithm (Fcm-Aco) for brain tumor segmentation. International Journal of Future Generation Communication and Networking 13(4): 3875–3886.

[21] Neha Taneja and Sahu, O. P. 2014. MRI brain tumor segmentation using improved ACO. International Journal of Electronic and Electrical Engineering, ISSN 0974-2174, 7(1): 85–90.

[22] Taiwo Oladipupo Ayodele. 2010. Machine learning overview. New Advances in Machine Learning, February 2010, pp. 9–18.

[23] Machine Learning, Tutorials Point. 2019. https://www.tutorialspoint.com/machine_learning/machine_learning_tutorial.pdf.

[24] Dervis Karaboga. 2010. Artificial bee colony algorithm. Scholarpedia 5(3): 6915.

[25] Karaboga, D. 2005. An Idea based on Bee Swarm for Numerical Optimisation. Technical Report-TR06, October 2005.

[26] Menon, N. and Ramakrishnan, R. 2015. Brain tumor segmentation in MRI images using unsupervised artificial bee colony algorithm and FCM clustering. 2015 International

Conference on Communications and Signal Processing (ICCSP). doi:10.1109/iccsp.2015.7322635.

[27] Mutasem Alsmadi. 2015. MRI brain image segmentation using a hybrid artificial bee colony algorithm with fuzzy–C mean algorithm. Journal of Applied Sciences 15(1): 100–109.

[28] Hancer, E., Ozturk, C. and Karaboga, D. 2013. Extraction of brain tumors from MRI images with artificial bee colony based segmentation methodology. 2013 8th International Conference on Electrical and Electronics Engineering (ELECO), pp. 516–520, doi: 10.1109/ELECO.2013.6713896.

[29] Hemanth, G., Janardhan, M. and Sujihelen, L. 2019. Design and implementing brain tumor detection using machine learning approach. 2019 3rd International Conference on Trends in Electronics and Informatics (ICOEI). doi:10.1109/icoei.2019.8862553.

[30] Ravikumar Gurusamy and Vijayan Subramaniam. 2017. A machine learning approach for MRI brain tumor classification. CMC 53(2): 91–108.

[31] Khalid Usman and Kashif Rajpoot. 2017. Brain tumor classification from multi-modality MRI using wavelets and machine learning. Pattern Anal. Applic. 20: 871–881. DOI 10.1007/s10044-017-0597-8.

[32] Pawel Mlynarski, Hervé Delingette, Antonio Criminisi and Nicholas Ayache. 2019. Deep learning with mixed supervision for brain tumor segmentation. Journal of Medical Imaging 6(3): 034002 (Jul–Sep 2019).

CHAPTER 8
Analysis of Machine Learning Algorithms for Prediction of Diabetes

Amartya Bhattacharya* and *Rupam Kumar Roy*

1. Introduction

Diabetes mellitus or diabetes is a type of disease which is caused by increased sugar concentration level in our blood, also known as blood sugar. Blood sugar, in general, serves an important role in developing energy in our body and the main source of this is the food which we consume daily. The sugar concentration in the blood is controlled by a substance known as insulin. This is a hormone produced by the pancreas in our body and is involved in the process of conversion of glucose obtained from the food, into a source of energy in our body. In some cases, due to various reasons enough insulin is not produced in our body thus leading to accumulation of glucose which in turn leads to diabetes.

Accumulation of glucose leads to the disease which can be prevented by following various steps and taking several measures.

The cases of diabetes in India are on the rise. Recent reports suggest that the cases rose from around 23 million in 1985 to around 60 million in 2015. More shocking facts are obtained from the National Diabetes and Diabetic survey report by the Ministry of Health and Family Healthcare, India, where it is stated that around 12% of the people having aged above 50 are affected with the disease. Also in a survey by the DHS, it is said that almost 7% of the

Department of Computer Science and Engineering, University of Calcutta.
Email: rupamkrroy@gmail.com
* Corresponding author: amartyab.ju@gmail.com

adults below the age of 50 are affected with diabetes while 6% are in a stage of having diabetes in the near future. Although the percentage of affected people is almost the same for both genders it has been found that people living in the urban areas are affected more than the people living in rural areas. Diabetes also leads to a certain condition known as diabetic retinopathy which affects the eyesight and can have major effects in future. The surveys have shown that the cases of diabetic retinopathy for the people having an age above 50, is around 17%. The report, more specifically shows that the cases of diabetic retinopathy are around 18.6% for the people belonging to the age group of 60–70 while 18.3% for people having an age between 70–80. Although the percentage is low for the age group of 50–60, i.e., around 14.3%, people having an age more than 80 years are mostly affected with surveys showing of around 18.4% cases in that age group.

One interesting fact suggests that the states of South India, mostly Tamil Nadu and Kerala, which perform better than the rest of the states in terms of economy and hospitality, also have a high number of cases for the disease which is alarming. Thus there is a huge need for early detection and predictions for the disease, which affects such a large number of population in India and worldwide.

In some of the previous works done by the researchers, prediction of diabetes has been done using different Machine learning classifiers but their accuracy was not so high and was around 75–77%. In our paper, we have analyzed the diabetic issue through a dataset [1] obtained from the UCI machine learning lab repository. The dataset is also present in Kaggle. The data of 768 female patients with a minimum age of 21 years of Pima Indian Heritage is contained in the dataset. In our study 7 different types of machine learning classifiers, i.e., Logistic Regression, K Nearest Neighbors, Classification Tree, Random Forest Classifier, XGBoost Classifier, Adaboost Classifier and LGBM Classifier have been used for prediction. In addition we have applied 10 fold cross validation with hyper parameter tuning using GridSearchCV and improved the result drastically and got the highest accuracy of around 87% with this method.

2. Methodology

Our target variable is an "Outcome" with 2 distinct results, i.e., whether a person has diabetes denoted by '1' or not, denoted by '0'. It was noted that the given outcome is of numeric type which suggested that we need not label encode the outcome, in order to train or predict.

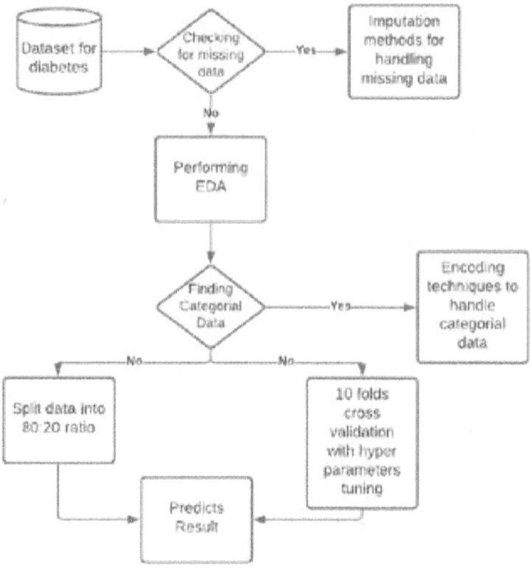

Fig. 1. Flow diagram of the paper.

The operations performed were - :

i) *Checking for missing data*: The dataset was analyzed using the mean of the isnull() function to check the presence of null or 'NaN' values but none of them were found.

ii) *Exploratory Data Analysis (EDA)*: We have performed EDA on our dataset to check whether any categorical data is present or not. Categorical variables were also not present in the given dataset thus preventing us from performing various preprocessing steps like encoding techniques.

iii) *Splitting the dataset*: Then we have split our dataset in two ways, first the dataset is split in the ratio, 80:20 where 80% data is used for training our models and 20% for predicting the result and another is to use 10 folds cross validation with hyper parameter tuning using GridSearchCV to get the optimum result.

At the end we have compared the predictive outcomes of different classifiers.

Some of the terminologies that we have used in our result section are-

i) *Accuracy*: The accuracy for classification problems denotes the ratio of the number of correct predictions both in cases of positive and negative and the total number of predictions considering both the false and the true.

ii) *ROC AUC Score*: ROC-AUC is a measure of performance for classification problems which shows the results by varying the threshold separating the classes. The ROC (Receiver Operator Characteristics) curve is generally obtained by plotting the probabilities and the area of the curve or the AUC helps us to know how well can the classes be distinguished with the help of the given algorithm. Generally higher value of AUC implies the model can distinguish the different classes in a better way, that is, correctly predicting the classes for example 0 class as 0 and 1 class as 1. So, we tried to compare the different algorithms by using the AUC scores.

3. Machine Learning Classifiers

i) *Logistic Regression* [2, 3]: Logistic regression is a linear classifier which helps in classifying a result into two classes based on the probability of the outcomes. Since the outcome variable had only two classes, this was used.

ii) *K Nearest Neighbor* [4, 5]: K-NN is an algorithm used for both classification and regression but here it is used for classification. It is used in tracking the previous classes of the data and classifying the new data based on the similarity of the classes of 'K' data points, nearest to the new da where TP represents the number of True Positives, TN represents the number of True Negatives, FP represents the number of False Positives and FN represents the False Negatives. For Hyperparameter tuning the values of 'K' were varied between 5, 7, 8 and 10 in order to note the 'K' value with the best results as, using values more than 10 led to the fall of accuracy The metrics for calculating the distance of the 'K' data points were also varied between 'Euclidean', 'Manhattan', 'Chebyshev' and 'Minkowski' distances in order to find the best metrics with the best results.

iii) *Classification Tree* [6, 7]: It is a supervised algorithm which can be used both for classification and regression. Here it is implemented for classifying the target variable (i.e., outcome) by forming nodes (tests), edges (checking whether the data satisfies the test or not) and leaves (which decides the outcome). In this problem the impurity of each node was estimated with both gini coefficient and entropy, and the maximum depth of the tree was selected from the range of 1 to 100, having the highest accuracy.

iv) *Random Forest Classifier* [8]: The Random Forest Classifier is a bagging algorithm which uses multiple Decision Trees in order to classify the data points into different classes. For hyper parameter tuning the estimators were chosen from between 100, 150 and 200 while the criterion for splitting the trees was checked both using Gini impurity and Entropy in order to find the one with the best results.

v) *Adaboost Classifier* [9, 10]: It is an algorithm for binary classification. It is done with the help of short decision trees using boosting techniques and assigning weights for each instance of the training set. The number of trees was varied from 100, 150 and 200 using hyper parameter tuning while the learning rate was varied from between 0.1, 0.5, 0.8 and 0.1 to get the best combination in terms of results.

vi) *XGBoost* [11]: It is a decision tree based ensemble algorithm. It uses a gradient boosting framework. The learning rate was varied from 0.01 to 0.1 in order to find the one with the best results.

vii) *LGBM Classifier*: Light Gradient Boost Machine Classifier or LGBM Classifier is a machine learning algorithm which uses the gradient boosting method, where the training algorithm is changed making the whole training process much faster than the other gradient boosting techniques. The learning rate was varied between 0.01, 0.05 and 0.1 in order to achieve the highest accuracy value.

4. About Dataset

The dataset is originally from National Institute of Diabetes and Digestive and Kidney Diseases and is present in the UCI Machine Learning Lab data repository as well as the Kaggle data repository. This dataset contains the data of 768 female patients with a minimum age of 21 years of Pima Indian Heritage.

The features present in the dataset in order to help predict the presence of diabetes are :

1. **Pregnancies:** The number of pregnant female patients in the dataset has gone through with a minimum value of 0 and maximum of 17 with an average of 3 pregnancies and 1 being the most frequent value.

2. **Glucose:** The plasma glucose concentration level for 2 hours found through an oral glucose concentration test. The values range from 0 to 199 with 120 being the average.

3. **Blood Pressure:** The systolic blood pressure level had values ranging from 0 to 122 with 69 being the average.

4. **Skin Thickness:** The skin fold in triceps was measured in 'mm' with values present in between 0 to 99, with a mean of 21.

5. **Insulin:** The 2 hour serum insulin level was measured in 'mu U/ml' with the values ranging from 0 to 846 with the average value being 80.

6. **BMI:** The body mass index of a person calculated by the formula weight in 'Kg' divided by the square of the height in 'm'. The value of BMI ranges from 0.0 to 67.1 with an average of 32.

7. **Diabetes Pedigree Function:** The diabetes pedigree function with values in the range of 0.078 to 2.42 with an average of 0.47.

8. **Age:** The age of the female patients present in the dataset measured in 'years' with minimum age of 21 years and maximum of 81 years with 33 years being the average.

The presence of diabetes in the patients present in the dataset is represented by an **Outcome** which has two values, i.e., 0 and 1 in almost a balanced distribution of 268 out of 768 patients have the value 1 which represents the presence of diabetes and the remaining 500 patients have the Outcome value of 0 which represents they don't have diabetes.

5. Result & Discussion

Initially we observe the results without using K fold cross validation and thus not tuning the hyper parameters.

The metrics which were considered were Accuracy which is given by equation 1.

Standard Deviation of Accuracy and ROC-AUC Score.

$$Accuracy = (TP + TN)/(TP + TN + FP + FN) \qquad (1)$$

where TP represents the number of True Positives, TN represents the number of True Negatives, FP represents the number of False Positives and FN represents the False Negatives obtained from the confusion matrix.

From Table 1. It can be inferred that Random forest classifier performed the best with XGBoost performing slightly less than it without cross validation.

After the initial observation we used hyperparameter tuning with 10 fold cross validation in order to note the best combination of hyperparameters in terms of the results. For the mentioned process we used GridSearchCV. Given below are the optimal combination of hyperparameters that gave the best results.

Table 1. Performance analysis of different ML classifiers.

Algorithms	Accuracy (%)	ROC AUC	Accuracy (%) with 10 Fold	ROC AUC With 10 Fold
Logistic Regression	77.92	0.71	78.61	0.85
K Nearest Neighbours	74.67	0.69	77.67	0.80
Classification Tree	70.77	0.67	71.87	0.67
Random Forest Classifier	82.46	0.78	81.16	0.86
Adaboost Classifier	74.67	0.71	75.87	0.83
XGBoost Classifier	81.16	0.79	87.14	0.87
LGBM Classifier	79.02	0.76	79.67	0.83

From Table 1 it can be observed that after applying 10 folds cross validations and hyper parameter tuning we got the best accuracy with XGBoost, which has an accuracy of 86.14%. After applying 10 folds cross validation random forest has performed well too with an accuracy of 81.16%.

Following are the confusion matrices and ROC AUC curves for different ML classifiers-

(i) Logistics Regression-

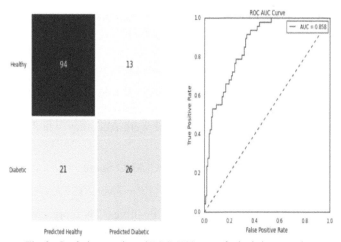

Fig. 2. Confusion matrix and ROC AUC curve for logistic regression.

(ii) KNN-

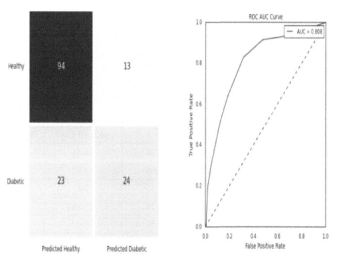

Fig. 3. Confusion matrix and ROC AUC curve for KNN.

(iii) Classification Tree

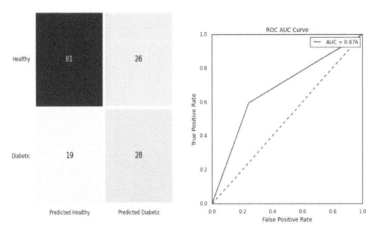

Fig. 4. Confusion matrix and ROC AUC curve for classification tree.

(iv) Random Forest Classifier

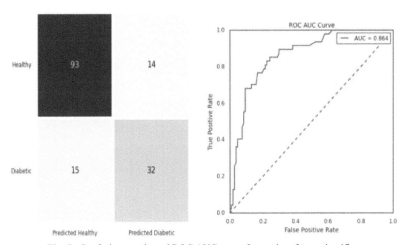

Fig. 5. Confusion matrix and ROC AUC curve for random forest classifier.

(v) Adaboost Classifier

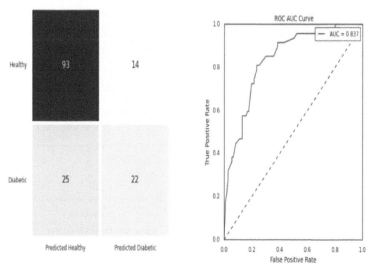

Fig. 6. Confusion matrix and ROC AUC curve for Adaboost classifier.

(vi) LGBM Classifier

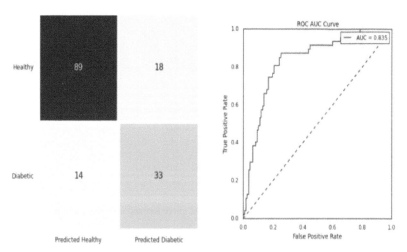

Fig. 7. Confusion matrix and ROC AUC curve for LGBM classifier.

(vii) XGBoost Classifier

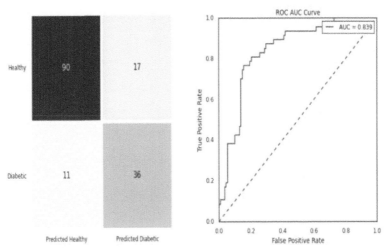

Fig. 8. Confusion matrix and ROC AUC curve for XGBoost classifier.

Variation of Accuracy Scores

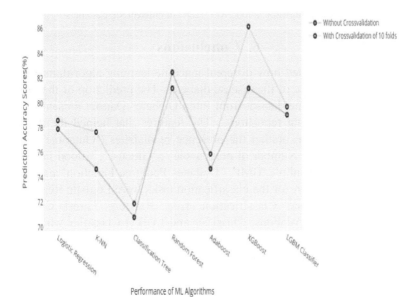

Fig. 9. Comparison of accuracy scores of different ML classifiers.

Variation of ROC-AUC Scores

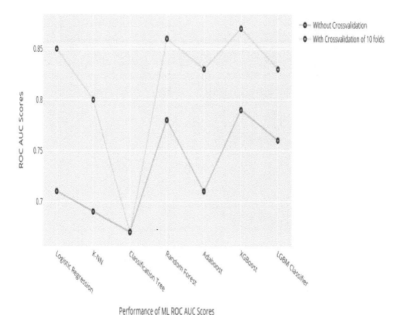

Performance of ML ROC AUC Scores

Fig. 10. Comparison of ROC AUC scores of different ML classifiers.

6. Conclusions

In our paper we studied how different machine learning algorithms help us to predict the presence of the disease diabetes. The prediction of the disease was done by taking the dataset from Pima Diabetes Dataset present in UCI Machine Learning data repository... The features that helped the machine learning algorithms to predict the presence of diabetes ('Outcome' feature in the dataset) were, 'Number of pregnancies', 'Glucose', 'Blood Pressure', 'Skin thickness', 'Insulin', 'BMI', 'Diabetes Pedigree Function' and 'Age'. The algorithms chosen for the classification tasks were Logistic Regression, K Nearest Neighbors, Classification Tree, Random Forest Classifier, XGBoost Classifier, Adaboost Classifier and LGBM Classifier which were first trained with the training dataset and the results obtained and reported were done by using cross validation, performed 10 times. The results obtained by various algorithms were then compared using the AUC score and Accuracy score. By comparing the different algorithms, it was noted that the XGBoost algorithm performed the best over the given dataset. However, the results are not absolute since we didn't consider any other dataset and the dataset under consideration was relatively small and also the instances were less.

References

[1] Smith, J. W., Everhart, J. E., Dickson, W. C., Knowler, W. C. et al. 1988. Using the ADAP learning algorithm to forecast the onset of diabetes mellitus. pp. 261–265. In Proceedings of the Symposium on Computer Applications and Medical Care. IEEE Computer Society Press.

[2] Wright, R. E. 1995. Logistic regression.

[3] Hosmer Jr, D. W., Lemeshow, S. and Sturdivant, R. X. 2013. Applied Logistic Regression (Vol. 398). John Wiley & Sons.

[4] Christobel, Y. A. and Sivaprakasam, P. 2013. A new class wise k nearest neighbor (cknn) method for the classification of diabetes dataset. International Journal of Engineering and Advanced Technology 2(3): 396–200.

[5] Pham, H. N. A. and Triantaphyllou, E. 2008. Prediction of diabetes by employing a new data mining approach which balances fitting and generalization. In Computer and Information Science. Springer, pp. 11–26.

[6] Al Jarullah and Asma, A. 2011. Decision tree discovery for the diagnosis of type II diabetes. 2011 International Conference on Innovations in Information Technology. IEEE, pp. 303–307.

[7] Chen, W., Chen, S., Zhang, H. and Wu, T. T. 2017. A hybrid prediction model for type 2diabetes using k-means and decision tree. In 2017 8th IEEE International Conference on Software Engineering and Service Science (ICSESS), pp. 386–390.

[8] Coppersmith, D., Hong, S. J. and Hosking, J. R. 1999. Partitioning nominal attributes in decision trees. Data Mining and Knowledge Discovery 3(2): 197–217.

[9] Hastie, T., Rosset, S., Zhu, J. and Zou, H. 2009. Multi-class adaboost. Statistics and its Interface 2(3): 349–360.

[10] Li, X., Wang, L. and Sung, E. 2008. AdaBoost with SVM-based component classifiers. Engineering Applications of Artificial Intelligence 21(5): 785–795.

[11] Chen, T., He, T., Benesty, M., Khotilovich, V., Tang, Y. et al. 2015. Xgboost: extreme gradient boosting. R Package Version 0.4-2, 1(4): 1–4.

Chapter 9
Swarm Intelligence for Diagnosis of Arrhythmia and Cardiac Stenosis

K Sujatha,[1,]* NPG Bhavani,[2] B Latha,[3] K Senthil Kumar,[4]
T Kavitha,[5] U Jayalatsumi,[6] A Kalaivani,[7] V Srividhya[8] and
B Rengammal Sankari[9]

1. Introduction

The recent trends in monitoring the health conditions accounts for the diversified integration of devices and techniques. A variety of wearable devices like wrist bands and smart watches have been introduced in the market. These gadgets acquire bio-signals and convert them to equivalent electrical signals which are often used to denote the measured quantity. The corresponding quantities measured in the form of bio-signals represent the functionality of various organs. These biological signals have a definite mathematical function and their behaviour can be analysed. The malfunctioning of the heart which is

[1] Professor, EEE Department, Dr. MGR Educational and Research Institute.
[2] Associate Professor, Department of ECE, Saveetha School of Engineering, SIMATS, Chennai, Tamil Nadu, India.
[3] Professor, Physics Department, Dr. MGR Educational and Research Institute.
[4] Professor, Department of ECE, Dr. MGR Educational and Research Institute.
[5] Associate Professor, Department of Civil Engineering, Dr. MGR Educational and Research Institute.
[6] Associate Professor, ECE Department, Dr. MGR Educational and Research Institute.
[7] Associate Professor, Department of CSE, Saveetha School of Engineering, SIMATS, Chennai, Tamil Nadu, India.
[8] Assistant Professor, EEE Department, Meenakshi College of Engineering.
[9] Assistant Professor, EEE Department, Dr. MGR Educational and Research Institute.
Email: sbreddy9999@gmail.com
* Corresponding author: sujathak73586@gmail.com

a vital organ of the human circulatory system should be viewed seriously as it will prove dangerous.

The heart in the human body is only one fist zone. It is the strongest muscle in the human body. The heart starts to beat in the uterus long before birth, usually by 21 to 28 days after conception. The average heart beats about 100000 times daily or about two and a half billion times over a 70 year lifetime. With every heartbeat, the heart pumps blood around the body. It beats approximately 70 times a minute, although this rate can double during exercise or at times of extreme emotion [3, 4].

From the left chamber blood is pumped out of the heart to all other parts in the human body. The arteries carry oxygenated blood to all parts of the body. The arteries finally branch as capillaries supplying oxygenated blood to all tissues, like skin and other major organs present in the human body. The deoxygenated blood after supplying oxygen and nutrients enters the heart for purification through the right chamber via the veins. When the oxygenated blood is circulated through the liver, the waste products are eliminated. Hence, if there is any breakdown in this network, the entire system collapses and may prove to be fatal to humans [5, 6]. When there is a hindrance in the blood vessels, there is a reduction in blood supply to the heart causing Coronary heart disease. The major factors contributing to coronary heart disease are high blood pressure, high saturated blood cholesterol, tobacco and alcohol use, unhealthy oily diet, physical inactivity, increase in blood sugar level and any other genetic disorders. The other factors include malnutrition due to poverty, lack of awareness, mental depression and swelling of blood vessels [7, 8].

Rheumatic heart disease damages the cardiac muscles and valves due to rheumatic fever, caused by streptococcal bacteria. Genetic factors contribute to congenital heart disease like hole in the heart, malfunctioning of heart valves and chambers. All these defects arise due to alcohol consumption, usage of medicines like warfarin during the pregnancy period, infections like rubella, malnutrition and consanguinity. Stroke occurs due to interrupted blood supply to the brain which may happen due to blockage (ischaemic stroke) or bursting of a blood vessel (haemorrhagic stroke). Deaths from cardiovascular diseases (CVDs) and advancing age prove to be challenge in the field of medicine [9].

2. Literature Survey

Statistical reports from the World Health Organization (WHO) state that the Cardio-Vascular Disease (CVD) contributed as a prime reason for death among human communities. There is a significant change in the ECG signals due to arrhythmia. By analyzing the ECG signals a non-intrusive technique can be proposed to identify Cardiac diseases. Physicians use a stethoscope to monitor the heart beat so as to identify the variations and detect the arrhythmia and cardiac stenosis [2]. This method may lag in effectiveness because the audibility range can be limited and is exposed to errors [3]. Analysis of an ECG signal is a non-invasive, bloodless method that represents the electrical activity of the heart [4, 5]. Any variation from the standard heart beat (72 beats per minute) is denoted as arrhythmia, inclusive of turbulence in the heart rate, reliability or transmission of the electrical impulses from the heart [6]. Recording the ECG signals for arrhythmia and cardiac stenosis detection usually requires a separate specialized equipment and medical unit along with physicians and biomedical engineers. Since, the fabrication of such devices is complex, costly and needs transportation, it will be a simple task to just use only the ECG equipment to capture the ECG signals and process them. This technology will be supported by all the nations worldwide, independent of their economic status as the proposed scheme incurs a lower cost. This system will be automatic, low price, real-time, and effective for detecting and monitoring the physiological parameters. All these conditions have created a favourable background for arrhythmia and cardiac stenosis detection using computer aided technology [10, 11].

Researchers have carried out the identification of heart disease using various machine learning algorithms. Heart disease occurs due to fat accumulation over a long period of time. The high pressure created within the human body creates heart disease. Various attributes were extracted from the data set to detect the cardiac disease. A dataset containing 12 parameters for 70000 samples from patients was used for analysis. Finally, a reasonable accuracy is achieved for detecting heart\disease using machine learning algorithms [10].

Blocks in the coronary artery are a common and frequently occurring heart disease causing heart attacks. Presently, Angiogram is used to diagnose blocks in coronary artery blockages and other heart diseases. Md. Ashraf ul Alam et al. [1], has used a fully computerized automated scheme for detection of blocks in the coronary artery using image processing techniques so as to avoid the manual intervention by a physician. Effective image processing and intelligent algorithms, facilitate a speedy and trustworthy detection of the reduced area of the wall of coronary arteries due to the condensation by blocking agents. Either a 64 or 128 slice CT image of the heart is used as input from which the region of interest is acquired. Two phases are involved, which include pre-processing and decision making. The pre-processing phase involves feature extraction which is used by AI algorithms to make a decision regarding the coronary heart disease. Hence early detection of coronary artery blockage is achieved by segmentation, quantification and identification of the degree of blockage causing heart attack [12, 13, 14].

The Angiography technique used for the detection of a coronary artery blockage leads to a variety of complications like artery dissection, arrhythmia and also death. Nayana Mohan et al. [2], have proposed an image processing method for identifying stenosis in regions of the artery. Digital Imaging and Communications in Medicine (DICOM) images of the heart are used for the detection of stenosis. The region of interest (coronary artery) is segmented using a vessel enhancement diffusion filter and morphological operations like dilation and erosion. The artery is segmented along the medial line using the fast marching based method. The diameter of the artery is measured along the medial line location in order to diagnose the blockage. The sudden decrease in the artery's diameter indicates stenosis. This method is evaluated for a set of images with an accuracy of 86.67% for diagnosing the stenosis [15, 16, 17].

The ECG signals which are bio-electric in nature are used for diagnosis of heart diseases. The proposed method of ECG analysis using CNN and RBFN can accurately detect Arrythmia conditions like Ventricular fibrillation, Atrial fibrillation and cardiac stenosis. The accuracy lies in the extraction of features from the ECG signals which is therefore a difficult job. The cardiologists from their experience can accurately forecast and recognize the correct heart disease from the ECG wave pattern and offer an appropriate treatment. This may involve observations and analysis of ECG signals of patients in critical care for a long period. So in order to automate the process a computerized diagnostic system has been developed to detect the abnormal conditions from the ECG wave patterns. The majority of the investigative studies in this area were conducted by incorporating different approaches of machine learning (ML) techniques for the resourceful recognition and precise testing of ECG signals [18, 19, 20].

The recorded ECG signals are extensively used for detecting and forecasting the cardiac stenosis and Arrhythmia conditions. Once the entire process is over the cardiologists will offer their diagnosis by referring to the ECG signal pattern, observed for a long duration to detect the cardiac arrhythmia and Stenosis. The ECG signals are one dimensional in nature which can be expressed as a time series function, analyzed using machine learning algorithms for cardiac stenosis detection and Arrhythmia conditions.

3. Materials and Methods

An artificial Intelligence (AI) research domain makes the machine perform tasks similar to human intelligence. Machine learning is used to attain skills without any human interference. Sophisticated intelligent machines can be deployed in the AI Systems which use artificial intelligence, machine learning and deep learning algorithms. Better decisions can be automatically made using Machine Learning algorithms [8]. A wide range of algorithms, on an iteration basis learns the data patterns and provides direction to forecast optimal outputs. Hence, the patterns are analyzed to identify suitable algorithms to detect the required output. The subset of machine learning is the Deep Learning neural networks. Optimal outcomes occur with human interference to achieve the desired output. They analyze the data in a logical structure to draw conclusions.

It is a part of machine learning which trains the computer about the basic instincts of a human being. The wider applications include virtual assistants, facial recognition and driverless cars among others. Supervised, semi-supervised and unsupervised learning algorithms are used to train the application related data. The following are the points to be considered for deep learning algorithms

- It mimics the activities of the human brain in decision making
- Patterns of data are varied, diversified and huge in number
- Decision making can be made efficiently.

Applying deep neural networks with multiple layers on a large volume of data compared to traditional Machine Learning algorithms is effective. Performance of deep learning algorithms is directly proportional to the amount and variety of data. These systems learn to take actions based on examples, without an explicitly specific program. Artificial Neural Network (ANN) architecture is made up of three layers—input, output and one hidden layer. Deep neural networks are ANNs with multiple layers between input and output layers which have multiple hidden layers. The major Deep Learning algorithms are Deep Neural Network, Deep belief Network, Recurrent Neural Network and Convolutional Neural Network. These algorithms are applied for different applications based on the requirement and performance with different types of data [13].

3.1 Classification of Deep Learning Algorithms

Deep neural networks are not easy to train with back propagation due to the problem of vanishing gradient which impacts the time taken for training and reduces accuracy. The cost function is calculated based on the net difference between the Neural Network's predicted output and actual output in the training data. Based on the cost, weights and biases are altered after each process until the cost is small. Gradient is the rate at which cost will change based on weights and biases. Deep learning algorithms perform best with problems with huge data sets.

3.2 Convolutional Neural Network (CNN)

A CNN consists of three layers as shown in Fig. 1. They are, Convolution, Pooling and fully connected layers. The convolution function is the integral product of the two functions after one is reversed and shifted (Reyes-Galaviz et al. 2008). The convolution layer is the first layer and it extracts the features from an input image. In the convolutional layer an activation function is applied which helps in getting non-linearity and an output based on the input [16]. The activation functions may be ReLU, Sigmoid, or tan h. The most commonly used in CNN is ReLU.

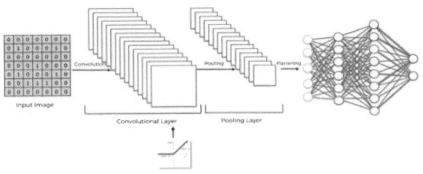

Fig. 1. General architecture of convolution neural network.

3.3 Objectives of the Work

The prime objective behind this work is the immediate detection of coronary artery stenosis and Arrhythmia conditions like Ventricular fibrillation and Atrial fibrillation from the ECG signals. As on date, the diagnosis is done in consultation with a Cardiologist by direct diagnosis to detect the coronary artery stenosis and Arrhythmia conditions like Ventricular fibrillation and Atrial fibrillation from the Angiogram by clinical examination. This is a time consuming procedure and also a challenging task. The Angiogram also has a disadvantage where there is a time gap involved in the analysis and increases the chance of risk due to infections. Smart health monitoring using intelligent image processing algorithms enables continuous, safe, remote and online diagnosis. The proposed scheme is a non-invasive method which uses, signal processing and intelligent algorithms for online detection of coronary artery stenosis and Arrhythmia conditions like Ventricular fibrillation and Atrial fibrillation from ECG signals. Thus, this indigenous technique once deployed will reduce the encumbrances faced by the patients who find it a challenge to immediately identify heart diseases from an Angiogram with reduction in cost needed for the development of a separate device or instrument thereby preventing the usage of very difficult and time-consuming technologies. Also, this technology can cater to the needs of a large number of heart patients by analysing their ECG signal enabling the maintenance of social distancing during this pandemic COVID-19.

i.   The proposed technology is a smart phone App for detection of Arrhythmia conditions like Ventricular fibrillation and Atrial fibrillation from the ECG signals

ii. This smart technology uses intelligent image processing algorithms like Discrete Fourier Transform (DFT) and Convolutional Neural Network (CNN) for classification of the ECG signal captured

iii. Time delay will be reduced, and results will be obtained in fast

iv. It will facilitate remote, online and continuous monitoring for detection of Arrhythmia conditions like Ventricular fibrillation and Atrial fibrillation from the ECG signals and will be able to analyze at least 70,000 samples within fast.

v.   This indigenous intelligent signal processing algorithm when developed will be able to cater to the need of many users thereby enabling social distancing during this pandemic COVID-19. The entire system also will have a cloud set which will facilitate data retrieval any time.

The novelty of the proposed online based smart system is the detection of Arrhythmia conditions like Ventricular fibrillation and Atrial fibrillation from the ECG signals. This painless method enables user friendly diagnosis for cardiologists by embedding intelligent signal processing algorithms which will process the ECG signals captured. This method uses the Power spectral values calculated from DFT which are applied to the ECG signals that are captured. The power spectral coefficients will classify the Arrhythmia conditions like Ventricular fibrillation and Atrial fibrillation from the ECG signals with specific and accurate measurements instantly. On the other dimension, this novel technology will place an end to the challenge involved, thereby offering a contactless diagnosis system during this pandemic Covid 19 situation.

3.4 Gaps Identified

Presently, Angiogram is used as the existing technique for detection of Arrhythmia conditions like Ventricular fibrillation and Atrial fibrillation. The researchers have not developed the technology to infer the heart related problems for heart patients by analysing the ECG signal. So far, ECG analysis is being used as a preliminary technique for detecting the heart related problems. Hence from our side, we have focused to develop a robust intelligent image processing algorithm which will use intelligent signal processing algorithms to detect the reasons for Arrhythmia conditions like Ventricular fibrillation and Atrial fibrillation from ECG signals.

It is understood that a very small quantum of work has been reported in this area, which serves as a strong support to carry out this work. This proposed technology, will develop smart algorithms which will capture the ECG signals, process them to determine the reasons for Arrhythmia conditions like Ventricular fibrillation and Atrial fibrillation.

The proposed technology which will infer the reasons for Arrhythmia conditions like Ventricular fibrillation and Atrial fibrillation from ECG signals which is an indigenous technique to relieve the anxiety of patients continuously suffering from analgesic pain. Also, this technology will reduce the time delay involved in contacting the cardiologist, thereby offering an immediate first aid. Moreover, the proposed technology is a layman approach which facilitates the patients to face a challenge to identify the exact reason at the initial stages itself.

3.5 Significance of the Proposed Technology

The proposed technology will find major applications in the following sectors

   i. Safe, secure and emergency online measurements for heart patients
  ii. Biomedical instrumentation Industries and Mobile health monitoring vehicles
 iii. Hospitals and their associated institutions for Medical studies
  iv. Research Centre and Utility in COVID-19 zones
   v. Nationalized laboratories
  vi. Engineering and Technology related Institutions
 vii. Diagnostic centres

4. Methodology

The ECG signals of the heart for 140 patients were reviewed for identification of blocks in the heart followed by arrhythmia detection. Firstly, 122 ECG signals were analyzed to identify and measure blocks in the arteries present in the human heart and 18 more ECG signals for validation. Randomly, the samples are selected with various percentages of blocks in the coronary artery. This meant that nearly, 42 patients had no coronary artery disease, second 40 patients had 10%–70% of coronary artery block and third 40 patients had greater than 70% of coronary artery block. The ECG signals shown in Fig. 2 corresponds to various levels of coronary artery blocks obtained from the open source database (https://openheart.bmj.com/).

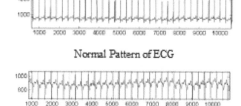

Normal Pattern of ECG

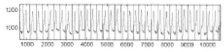

Pattern of ECG for patients with Arrhythmia conditions like Ventricular fibrillation

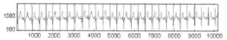

Pattern of ECG for patients with Arrhythmia conditions like Atrial fibrillation

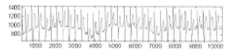

Pattern of ECG for patients with coronary artery block (10-50%)

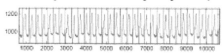

Pattern of ECG for patients with coronary artery block (50%-70%)

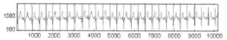

Pattern of ECG for patients with coronary artery block = 70%

Fig. 2. ECG signals.

The novel CNN classifier was realized in MATLAB. Extensive computation has taken place during the CNN architecture training. The investigation was done using a super high speed CPU. The entire data set was segregated as ECG signal patterns which were separated in such a way that 75% of the data was used for training and 25% was used for testing. A 3-fold cross validation was carried out during the training of the CNN.

The principle of operation of the customized smart measurements for analysis of ECG signals to detect Arrhythmia conditions like Ventricular fibrillation and Atrial fibrillation and cardiac stenosis from ECG signals is discussed here. The proposed algorithm will have high accuracy with excellent screening capacities. Once the smart technology is developed, then the calibration will be done by asking the heart patients to use the smart technology. After that, the patient needs to present the ECG signals which will be directly sent to the processing units which will enable the algorithms to be executed with the given input ECG signals. The ECG signals are captured and then processed by the intelligent signal processing algorithms. The signal processing algorithms which are in built now, will select the region of interest (RoI). ECG signals that are captured are the actual inputs for this technology. Any noise due to interference will be automatically rejected using the DFT. For this purpose, only, Convolutional Neural Networks (CNN) will be used as classifiers which have the capacity to generalize and robustly detect the Ventricular fibrillation, Atrial fibrillation and cardiac stenosis there by preserving the accuracy level close to the standard level. When the ECG signal is captured, it will identify the RoI, present in the ECG signal and the power spectral coefficients details and approximations obtained from the DFT that lie outside the threshold which will be barred from detecting Ventricular fibrillation, Atrial fibrillation and cardiac stenosis. Once the technology is developed and installed, the results can be compared for measurements of various ECG signals for conditions like Ventricular fibrillation, Atrial fibrillation and cardiac stenosis by using CNN and RBFN classifier which use the convolution operation as the foundation that exists in the field of signal processing. These measurements pertaining to the conditions of the ECG signals like Ventricular fibrillation, Atrial fibrillation and cardiac stenosis can be transferred to cloud-based server for remote monitoring rather than on-board examinations. The schematic is depicted in Fig. 3.

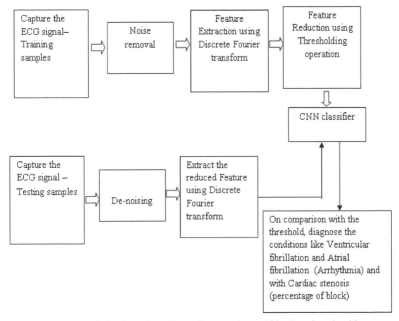

Fig. 3. Schematic for detection of heart disease using machine learning algorithms.

4.1 Pre-processing

The noise removal is also done using the DFT. De-noising helps to improve the quality of the ECG signal, so that appropriate and exact values of the features alone can be extracted. A High Pass Filter (HPF) allows the signal portion corresponding to high frequency values alone to pass through it. The transfer function for the HPF in a discrete domain is given by $1-0.77z^{-1}$.

The ECG signals comprise of three types of noise. The first one is the interference in power lines, the second one is the baseline drift and third there is an electromyographic noise. Noise free ECG signals must be used to de-noise the ECG signals. For this purpose DFT with an appropriate threshold is used. The DFT along with simple thresholding helps to reduce the electromyographic noise and noise due to power line interference. The baseline drift can be removed to eliminate the noise in ECG signal.

4.2 Feature Extraction using Discrete Fourier Transform (DFT)

An extensive variety of features are extracted from the ECG signal using a one dimensional discrete wavelet transform as illustrated in Fig. 4. The preferred feature set serves as the input and includes the details and approximation components to train and test the CNN.

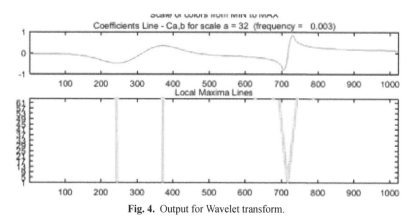

Fig. 4. Output for Wavelet transform.

4.3 Identification using CNN

A supervisory ECG signal segmentation scheme, to trace and capture the Ventricular fibrillation, Atrial fibrillation and cardiac stenosis even in a noisy environment can be facilitated by implementing the CNN to detect the abnormality in the training phase by using 70% of the collected signals itself, so that an appropriate model for cry signal analysis can be launched. During testing of CNN, only 20% of the collected sample signals are used for detection of abnormality. The remaining 10% can be used for validation after the development of the proposed technology.

5. Results and Discussion

The detection of blockages in the coronary artery at various levels is substantiated with quantitative results for detection supported by training and validation of the data collected. At the detection stage by the CNN and RBFN, a sensitivity of 97% and a positive predictive value (PPV) of 96% is achieved for blocks > 70%, a sensitivity of 96% and a positive predictive value (PPV) of 94% is achieved for a block in the range of 50%–70%, a sensitivity of 92% and a positive predictive value (PPV) of 93% is achieved for a block in the range of 10%–50%. These results are compared to BPA, while a sensitivity of 18% and a PPV of 24% for blocks > 70%, a sensitivity of 16% and a positive predictive value (PPV) of 22% is achieved for a block in the range of 50%–70%, a sensitivity of 12% and a positive predictive value (PPV) of 23% is achieved for a block in the range of 10%–50%, are obtained as compared to RBFN. However, the blockage in the coronary artery is quantified with an Averaged Absolute Difference (AAD) of only 2.7% for CNN as compared to RBFN. Thus, although the algorithm performed relatively well for detection of 50% constriction in coronary artery, it remains to be improved for better categorization of the block identification with mild, moderate, severe and occlusion categories. Finally, the average execution time for CNN and RBFN of the ECG signals with 3.4 GHz processor is just about 9.9 minutes.

The algorithm presented here is capable of a new automated system for detection of coronary artery stenosis and quantification of the blocks. An acceptable PPV was obtained with slightly less sensitivity on quantitative evaluation. The user interaction and detection time can be improved by making small changes with respect to the parameters in the algorithm so that the identification accuracy is maximized with reduction in AAD. Figures 5 and 6 show the results for training and Validation by CNN for various conditions of block in the coronary artery.

The efficiency of the CNN algorithm was compared with RBFN to detect arrhythmia conditions like Ventricular fibrillation, Atrial fibrillation and cardiac stenosis by the classification technique. The usual heart beat for normal conditions (NOR) and additionally two more types of cardiac arrhythmia like Ventricular fibrillation, and Atrial fibrillation identified from the ECG signals. From the entire set of representations the two types of cardiac arrhythmia like Ventricular fibrillation, and Atrial fibrillation are taken into account. Similarly, normal conditions (NOR) and cardiac stenosis were also identified.

The objective function is used to calculate the error of the proposed CNN architecture between actual and target values. The optimization technique

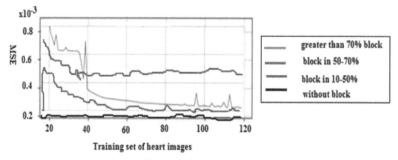

Fig. 5. Training of RBFN.

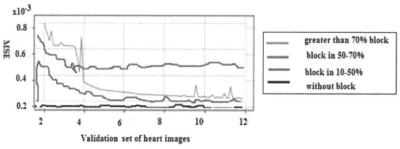

Fig. 6. Validation of RBFN.

is an optimizer function used to minimize the error function. Different cost functions have been used in the neural network theory. In our experiments, we used the cross-entropy. The model follows the CNN architecture with four 2-D convolutional layers. Each convolutional layer is followed by a pooling layer. The output layer is a SoftMax layer with eight neurons to give the final classification. A fully connected layer is used between the last pooling layer and the output layer and represents the features learned by the CNN model.

6. Conclusion

In this research paper, a CNN and RBFN based classification method is presented to detect the coronary artery stenosis and Arrhythmia conditions like Ventricular fibrillation and Atrial fibrillation from the ECG signals. It is inferred that a CNN classifier is more robust as compared with RBFN, so as to reduce the false positives maintaining a high accuracy in the identification rate. At the first stage, the method discriminates the arterial block followed by the density of blocks present within the regions of the heart. This method also helps to diagnose the Arrhythmia conditions like Ventricular fibrillation and Atrial fibrillation from the ECG signals. Thus, a classification based approach has the potential to ease the difficulty of the physicians by reducing the rate of manual interactions which is eventually required to, eliminate them completely.

References

[1] Md. Ashraful Alam, Mohsinul Bari Shakir and Monirul Islam Pavel. 2019. Early detection of coronary artery blockage using image processing: segmentation, quantification, identification of degree of blockage and risk factors of heart attack. Proceedings Volume 10982, Micro- and Nanotechnology Sensors, Systems, and Applications XI.

[2] Nayana Mohan and Vishnukumar, S. 2016. Detection and localization of coronary artery stenotic segments using image processing. International Conference on Emerging Technological Trends, IEEE Xplore.

[3] Oksuz, D. and Unay, K. Kadipasaoglu. 2012. A hybrid method for coronary artery stenosis detection and quantification in CTA images. Funded Project of Turkish Ministry of Science Industry and Technology.

[4] Kurkure Uday, Deepak R. Chittajallu, Gerd Brunner, Yen H. Le and Ioannis A. Kakadiaris, 2010. A supervised classification-based method for coronary calcium detection in non-contrast CT. The International Journal of Cardiovascular Imaging 26(7): 817–828.

[5] Shahzad, R., Michiel Schaap, Frederico Bastos Goncalves, Metz, C. T., Hui Tang, Theo Van Walsum, Adriaan Moelker, Lucas J. Van Vliet and Wiro J. Niessen. 2012. Automatic detection of calcified lesions in the descending aorta using contrast enhanced CT scans. Biomedical Imaging (ISBI) 2012 9th IEEE International Symposium on, pp. 250–253.

[6] Martijn S. Dirksen, Hildo J. Lamb, Joost Doornbos, Jeroen J. Bax, Wouter Jukema, J. and Albert De Roos. 2003. Coronary magnetic resonance angiography: technical developments and clinical applications: CORONARY ANGIOGRAPHY. Journal of Cardiovascular Magnetic Resonance 5(2): 365–386.

[7] Pankaj Goyal, Kuldeep Goyal and Vipin Gupta. 2013. Calcification detection in coronary arteries using image processing. Lancet. 3(2277): 599–603.

[8] Isgum Ivana, Mathias Prokop, Meindert Niemeijer, Max A. Viergever and Bram Van Ginneken. 2012. Automatic coronary calcium scoring in low-dose chest computed tomography. Medical Imaging IEEE Transactions on, 31(12): 2322–2334.

[9] Westin Carl-Fredrik, Lars Wigström, Tomas Loock, Lars Sjöqvist, Ron Kikinis and Hans Knutsson. 2001. Three-dimensional adaptive filtering in magnetic resonance angiography. Journal of Magnetic Resonance Imaging 14(1): 63–71.

[10] Chithambaram, T., Logesh Kannan, N. and Gowsalya, M. 2020. Heart disease detection using machine learning. Research Square.

[11] Animesh Hazra, Subrata Kumar Mandal, Amit Gupta and Arkomita Mukherjee. 2017. Heart disease diagnosis and prediction using machine learning and data mining techniques. Advances in Computational Sciences and Technology ISSN 0973-6107 10(7): 2137–2159. http://www.republication.

[12] Praveen Kumar Reddy, Sunil Kumar Reddy, T., Balakrishnan, Syed Muzamil Basha and Ravi Kumar Poluru. August 2019. Heart disease prediction using machine learning algorithm. Blue Eyes Intelligence Engineering & Sciences Publication. https:doi.org/10.35940/ijitee.J9340.0881019.com.

[13] Taylor, E., Ezekiel, P. S. and Deedam-Okuchaba, F. B. 2019. A model to detect heart disease using machine learning algorithm. International Journal of Computer Sciences and Engineering E-ISSN:2347-2693. 7(11): nov 2019. https://doi.org/10.26438/ijcse/v7i11.15.

[14] Nashif, Md. R. Raihan, Md. R. Islam and Imam, M. H. 2018. Heart disease detection by using machine learning algorithms and a real-time cardiovascular health monitoring system. World Journal of Engineering and Technology 6: 854–873. https://doi.org/10.4236/wjet.2018.64057.

[15] Svetlana Ulianova. 2019. Cardiovascular Disease dataset. The dataset consists of 70000 records of patient data, 11 features + target.

[16] Sonam Nikhar and Karandikar, A. M. 2016. Prediction of heart disease using machine learning algorithms. In International Journal of Advanced Engineering, Management and Science (IJAEMS) June 2016 vol-2.

[17] Deeanna Kelley. 2014. Heart disease: causes, prevention, and current research. In JCCC Honors Journal.

[18] Nabil Alshurafa, Costas Sideris, Mohammad Pourhomayoun, Haik Kalantarian and Majid Sarrafzadeh. 2017. Remote health monitoring outcome success prediction using baseline and first month intervention data. In IEEE Journal of Biomedical and Health Informatics.

[19] Ponrathi Athilingam, Bradlee Jenkins, Marcia Johansson and Miguel Labrador. 2017. A mobile health intervention to improve self-care in patients with heart failure: pilot randomized control trial. In JMIR Cardio 1(2): 1.

[20] DhafarHamed, Jwan K. Alwan, Mohamed Ibrahim and Mohammad B. Naeem. 2017. The utilisation of machine learning approaches for medical data classification. In Annual Conference on New Trends in Information & Communications Technology Applications - march 2017.

CHAPTER 10

Design and Development of Covid-19 Pandemic Situation-based Contactless Automated Teller Machine Operations

P Sivaram,[1,*] *Nandhagopal SM,*[2] *C Devaraj Verma*[3] and *Durga Shankar Baggam*[4]

1. Introduction

The gathering of FICs (people) is unavoidable when operating an ATM. The ATM operations are touch-based. Electronic(e) pay gateways are used in financial institutions (FI) for various cash transactions. Due to the pandemic, while it is beneficial for an FIC to avoid manual ATM operations, all FICs cannot use the latest technology devices to execute FI operations in remote modes, such as mobile banking, e-fund transfer. Further confidence of the users is low in e-financial transactions due to apprehension of cybercrimes. Nowadays, the conventional ATM operation(s) (C-ATM-O) are adopted by the FICs for transactions. The knowledge to use the system is easily available to the FICs. The stakeholders involved in such e-transactions must know fund transfer operations.

---

[1] Assistant Professor, Department of Computer Science and Engineering, School of Engineering & Technology, Jain University, Bangalore, India.
[2] Professor, Department of Computer Science and Engineering, Adithya Institute of Technology, Coimbatore, Tamilnadu, India.
[3] Associate Professor, Department of Computer Science and Engineering, School of Engineering and Technology, Jain University, Bangalore, India.
[4] Associate Professor, Department of Computer Science and Engineering, Gandhi Engineering College, Odisha, India.
* Corresponding author: ponsivs@yahoo.com

The proposed work focuses on using the existing C-ATM-Os. It progresses by introducing a contactless automated subsystem-based teller machine (CLASS-TM) framework of operations, an AIS hardware architecture with the flow of operations, and the necessary algorithms to govern the intended purpose. The article includes the necessary literature review needed to fulfill this work, the proposed discussion with the framework of operations, a hardware architecture, the supportive algorithms, the advantages of C19PCL-ATM, and future enhancements conclusion.

2. Literature Review

The available transactions of ATMs provided to the FICs are listed; these are time constraint-based processing. Access to ATMs is restricted based on time. The system's security is ensured with a timed automaton model [1]. The unique sign of hand vein available in each human is taken as a biometric factor for authentication, and recognition is accomplished by the neural network concept [2]. A multifactor authentication scheme was proposed to describe the ATM and voting process authentication. The optimum process outcomes were obtained from AI and Aadhar card-based unique systems [3]. As for the CL-ATM operations, FIC faces are recognized. Additionally, followed by an OTP-based authentication, the ATM security is measured and validated to proceed further with the ATM transactions [4]. For the rural and remote areas, FICs are provided with an e-door step banking system for the withdrawal and deposit operations of the bank with a handheld device in the form of a mini banking system [5]. The handheld machine is efficient in handling both cash withdrawal and deposit operations. With the newly developed Android app, the ATMs' status is tracked. The ATM facilities like cash dispense, out-of-order, and maintenance are provided in advance to the banking professionals and the FICs to avoid delayed banking services. The system ensures the FIC's interest in handy app-based updates rather than the traditional banking model [6]. With two main key factors, availability and reliability of ATM status, Amit Kumar and Varsha proposed a concept of various failures and reported in their work, how to avoid such failures [7]. The avoidance of the ATM card for the financial transaction need is discussed in the work of Imran et al. and enriched their work with cardless transactions and modelled a system based on bank ID number and an OTP that facilities the banking transaction, which is like CL-ATM operations [8]. Through the manual operation, the teller in the bank carries out the fund deposit and withdrawal operations. With the quick response code-based system with the FICs mobile app generated QR code, the communication between the teller and the FIC happens. The time-consuming manual verification and validation of the account by the teller is considerably reduced. The reduction in the time taken in the communication and fast processing is proposed as the outcome of the work done by Olayeye Ilemabayo et al. [9].

Biset Amene et al. carried out a study to investigate the influence of ATMs on customers. The investigated data is analyzed using data mining techniques. This study reveals that most persons use ATMs for cash withdrawal only and are mostly satisfied with services [10]. Swiecka et al. studied the effect of transaction factors on customers' payment choices. They built a system that studied customer behavior and predicted payment choices. Further, the customers' behavioral data is used to design financial policies and make strategic decisions [11]. An ATM service model with AI Technology introduced by Leonov et al. is the innovative work carried out with the AI subsystem [12].

Y. Najmi et al. described users' security, confidence, and reliability measures on IoT systems and presented a survey on security attacks. It helps the proposed system to deal with security attacks and determines such attacks [13]. Jain and Ranjan predicted the effects of emerging technologies on new application systems development and detailed the benefits of the inclusion of emerging technologies [14]. The work presented by S. Adewale et al. as a model for cardless ATM transactions provides our CL-ATM operations with high possibilities [15].

The effect on the bacterial infections due to ATM keypads used by FICs for their ATM operations are considered in this work. It was a measure of human diseases as the concept of this proposed work and was described by Iyabo A. Simon-Oke in his work and concluded that this type of ATM operation leads to a chance of risks in health conditions of FICs [16]. Jillet Sam et al. presented contactless cash operations and focused on payments due to Covid-19, and this work provides an open door for contactless ATM operations during the C19P situation [17]. This proposed work of C19PCL-ATM operations requires training of the FICs on the usage of highly secured ATM operations, presented by Benyam Tadesse and Fayera Bakala, for client satisfaction in ATM services [18]. The secured ATM transactions with GSM and biometric sensors based on ATM operations on a smart ATM provide a chance to the proposed system, which should be considered with secured operations and discussed in Balamurugan et al. [19]. The model presented by Yunhui Zeng et al. can be used with systems involved in HMI and background processing by are distinctly considered to model this article's proposed work [20]. Arun Singh et al. provided a biometric feature-based secure authentication for ATM access, which is consistent with the planned effort to improve the security of CL-ATM authentication. A computer vision-based camera scans the face of the FIC, and the authentication to access the ATM is provided with CL-ATM operation [21].

For FICs facing an authentication entity with a hand-free user interface PL. Chithra and B. Ilakkiya Arasi made a suggestion supporting the proposed C19PCL-ATM operations that are possible with the intended purpose [22]. Abiodun Esther Omolara et al. proposed that even people with visual impairments can play the role of FICs, using distinct touch-less ATM operations with IRIS recognition-based processes for transactions, which is very encouraging work for the proposed C19P CL-ATM [23]. In the current banking operations, existing static system-based controls are discussed and presented. Automatic intelligence based optimized banking operations with the introduction of AI subsystem-based machine learning provide possibilities to implement operations with dynamic authorization by B. Tay and A. Mourad is the main intention of our proposed work [24]. The online secured banking transactions are considered in the form of various authentications with the FIC's account. The threats against such transactions are addressed as the awareness provisions, which promote this proposed work [25]. Ameh Innocent Ameh et al. discussed the secured transactions of cardless ATMs with biometric and PIN authentication as a two-factor point of view [26]. Afia Farzana et al. presented secure transactions on ATM utilization with secure authentication, and C19PCL-ATM work designed with two-step authentication to provide secure CL-ATM operations [27]. For the C19PCL-ATM operations, it is considered that security is the crucial factor, where the transactions are subject to the monetary amount. Aleksander Biberaj et al. detailed the available logical threats of the ATMs and presented the possibilities of awareness to lead further development in the banking and ATM operations with security [28].

3. The Proposed C19PCL-ATM System

The provision of the SD-based ATM operations proposes to comply with the safety measures of the C19P situation. The proposed system upgrades the existing contact-based ATM operations into a contactless operating ATM by integrating them with available technologies. The proposed system incorporates an artificial subsystem in the existing ATMs to enable touchless user operations. The AI subsystems' components are given in Fig. 1.

Developers use the AI agents in various existing application systems, and modifications include the concept of AI-equipped system modernization [12]. In this scenario, the AI agents are equipped with their sensors and actuators for the environment operations. In the proposed system, the computer vision-based camera at the ATM doorstep is the sensor, a microphone is available on the ATM, and the CL-ATM app is on the user's mobile phone. The actuator is the electrical motor used to open and close the ATM's door. The ATM and its speakers act upon the customer's request for withdrawal or deposit operations. In the proposed work, these two primary operations were considered for

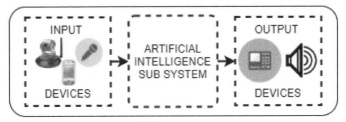

Fig. 1. Artificial subsystem components.

developing C19PCL-ATM operations. The regular operations of the ATM, like transaction alert through SMS, are provided to the customer's mobile number, as is happening with the existing ATMs.

The overall environment of C19PCL-ATM operations is the set of interconnected components to deliver and accomplish the scheduled task with the help of existing and proposed architecture. The existing system is the ATM network (ATM-N) of various financial institution(s) (FIs) with the highly secured concept of network architecture, which functions with a maximum throughput of its intended purpose and the ATM-Ns are optimum in excellence. The proposed system and its purposes are to provide the users with contactless operations due to the C19P situations of the entire world. The SD behavior in ATM access is to be incorporated with the existing ATM-N system. Figure 2 details the environment of the proposed C19PCL-ATM operations.

This proposed work's objective was to incorporate the AI subsystem (AI-SS) in the existing ATM-Ns. The FICs' mobile numbers are integrated

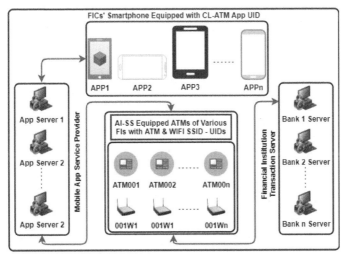

Fig. 2. C19PCL-ATM environment.

with their bank accounts for communication and act as a unique identifier (UID) in the existing FI system. The same UID is taken into account for the backend subsystem of that mobile app to identify FICs. The main components of C19PCL-ATM are (i) FIC's smartphone equipped with the CL-ATM app, (ii) the mobile app service provider's distributed networked servers, (iii) ATM units equipped with a WIFI facility and (iv) WIFI SSID's UIDs pairs, and the existing ATM-N's FI-transaction servers. The components (i) to (iii) are the proposed system modified components. The AI-SS is first incorporated at the mobile app level algorithm for FIC identification and authentication. The components (i) and (ii) are filled with AI algorithms to process and fix the appropriate FIC for the CL-ATM operations, and these algorithms are placed in the (ii) component of the C19PCL-ATM. The second AI-SS and its algorithms for facial recognition with a computer vision concept are placed on the (iii) component; the sensors are high pixel cameras at the ATM doorstep. The actuators are the electrical motors to open and close the ATM room door. With the AI-SS, the mobile app-based FIC executes a one-step authentication, and with the second AI-SS, facial recognition becomes the second step authentication. Thus, the two steps equip the ATM to authenticate reliably in a touch-free mode instead of the card and its touch-based PIN authentication. The environment of the C19PCl-ATM with its four components fulfils the intended purpose of the proposed work. Most of the proposed work design and development is in the mobile app service provider's distributed networked servers. The AI-SSs with two numbers are deployed for the effective functioning of the C19PCL-ATM. The first AI-SS is used to initialize CL-ATM in synchronizing the FIC's mobile app with the ATM with the help of the ATM's WIFI zone and acts as the first step of authentication with the FI system. The second AI-SS is used for authentication through facial recognition with the FI system and completes the two-factor authentication of the FI transactions with the fund distribution system (FDS)/ATM.

3.1 *C19PCL-ATM Operational IAAPO Framework*

The operation framework of C19PCL-ATM is categorized into five divisions: initialization, authentication, activation, processing, and operation commit (IAAPO). The IAAPO framework of C19PCL-ATM operations is finite and falls into completeness with the two AI-SSs introductions of the existing ATM-Ns. Figure 3 depicts the IAAPO framework of operations.

3.2 *Initialization*

The mobile app service provider is the third party that provides the mobile app distributed servers. The unique mobile app carries the concept of auto-running on the kernel as background execution during the mobile phone startup and

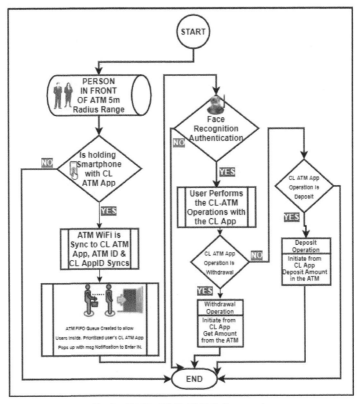

Fig. 3. C19PL-ATM's IAAPO framework.

waits for the WIFI state to auto access. The user with the mobile app, nearing a specific WIFI zone of the ATM, i.e., within five meters of the ATM, the SSID of the ATM WIFI signal is accessed by the mobile app. The ATM WIFI's SSID is predefined on the mobile app that initializes the operational execution in the mobile phone and gets connected with the ATM WIFI automatically with the first AI-SS algorithm. It is the initial stage of the mobile app synchronization with the ATM environment with the help of mobile app UIDs and the ATM zone's WIFI SSID pairs. The mobile app server has the priority queue of the FIC's synchronized with the ATM with a five meter range. The FIC's are provided with the order number to enter the ATM and get access to it with CL to perform the two operations (money deposit and withdrawal).

3.3 Authentication

The first AI-SS has operations up to the one-step authentication, i.e., mobile app synchronization with the ATM environment. The second step authentication with face recognition held at the priority-based FIC permits entry to the ATM.

The camera at the ATM doorstep acts as a vision-based sensor that captures the FIC's face data. The mobile app distributed network servers contain the face data of all the FICs to match the facial data captured by the ATM camera. Equality-based authentication is employed at this stage to allow the FIC to enter the ATM room with an AI-based ATM room door opening and closing mechanism. The two factors with two-stage authentication improve the ATM security of the conventional ATM-N's operations. Figure 4 depicts the facial recognition-based FIC's entry into the ATM.

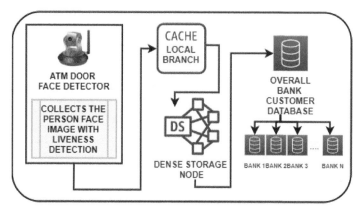

Fig. 4. Face recognition-based FIC's authentication for C19PCL-ATM operations.

3.4 Activation

After the two-step authentication, the mobile app server confirms the clarity of operations with the ATM and reports to the ATM that the FIC is an authenticated user. The conventional ATM operations are approved with this activation stage. The mobile app selects FIs from the FIC's available banks (linked with the mobile number as UID) with its pre-developed modules. Furthermore, the ATM money withdrawal and deposit operations are activated as the FIC's choice selection in the available FI accounts and selection of money deposit or withdrawal operations.

3.5 Processing

This operational framework allows the FICs to operate two ATM operations of their choice, a single operation at a time with perfect SD followers. With the face recognition of the FIC at the ATM doorstep camera, the FIC is identified, verified, and allowed to get into the ATM room. When the user is one meter away from the ATM, the mobile app pops up with the FIC's bank account selection options. The process further expands with the priority queue of FIC's ATM access. From the mobile app server-based priority queue allocation, the FIC receives the alert to enter the ATM.

The FICs may have more than one bank account registered on their mobile number, and the choice of selection is made with the initiated pop-up window of the mobile app. After this choice selection, the two ATM operations permitted with the proposed mobile app, cash withdrawal and cash deposit pop-up on the mobile app screen. Again, the user's choice-based selection of ATM operation initializes the appropriate ATM operation, and the conventional ATM operations are executed immediately after selecting choice operations with the mobile app. Figure 5 represents the processing framework operation of C19PCL-ATM, which is the proposed system's complete operation.

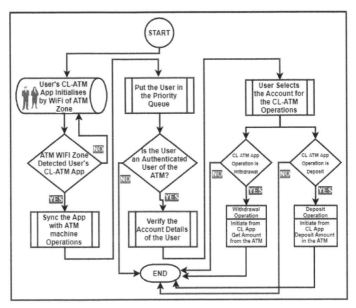

Fig. 5. Activated mobile app operations with C19PCL-ATM.

3.6 Operation Commit

Like the conventional ATM, the C19PCL-ATM allows the FIC to perform a single operation. The choice of selection of depositing or withdrawing an amount from the ATM rests with the FIC. If the FIC opts for withdrawal, the mobile app presents the details (Current Account/Savings Account, Amount to Withdraw, and Confirmation activity for Withdrawal) to perform a withdrawal and sends the request information to the ATM. The ATM further verifies the available balance of the FIC's account and the denomination of cash notes on the ATM and provides the conventional ATM services. In this stage, the withdrawal amount is dispensed in a sanitized withdrawal envelope and placed on the ATM tray designed to provide a specialty to the C19PCL-ATM functionality. The withdrawal operation is committed on the FI transaction servers as an update.

If the FIC opts to make a deposit, the mobile app represents the denomination of the cash notes option to be filled by the FIC and sends this information to the ATM. The ATM gets ready to collect the cash through its collection tray. At this stage, the deposit amount is absorbed by the ATM within its collection tray. The FICs are place the cash on the tray provided by the ATM to deposit the cash. The deposit cash collection mechanism specially designed with C19PCL-ATM collects the cash and verifies the denominations entered at the initial stage of the transaction. After successful verification, it sends a message confirming a successful cash deposit to the mobile app, which finishes the conventional ATM deposit operation. After this deposit operation, the FI transaction server is committed with the FIC's account's add-in amount as an update.

4. C19PCL-ATM Operational AI-SS Algorithms

The Mobile App server is equipped with the AI-SS1 and AI-SS2 algorithms for the C19PCL-ATM operations. The C19PCL-ATM Mobile App Server algorithm, i.e., AI-SS1, for the Mobile App to identify the ATM, initializes its operations, as shown in Fig. 6.

```
FUNCTION GPS_COORDINATES_MATCHING (LATITUDE, LONGITUDE)

Pre-Condition: CL-ATM_App running passively on the FIC's Mobile Device, waits for the
               matching of GPS_COORDINATES of Mobile Device with ATM

IF (GPS_COORDINATES_MOBILE ≈ ATM_GPS_COORDINATES)
   {
   MOBILEAPP_ACTIVE_STATUS (ON (1), OFF (0))
   IF (MOBILEAPP_ACTIVE_STATUS = 1)
      {
      INITIALIZE THE WIFI STATE OF MOBILE INTO ACTIVE.
      CHECK THE SSID OF THE ATM WIFI, PREDEFINED IN THE MOBILE APP.
      IF (MOBILEAPPDEFINED_SSID_VALUE = ATM_WIFI_SSID_VALUE)
         {
         AUTO CONNECTION ESTABLISHED BETWEEN MOBILE DEVICE TO THE ATM WIFI ACCESS.
         A TOKEN IS ASSIGNED FOR THE ATM ACCESS.
         THE MOBILEAPP FIC IS PLACED IN A PRIORITY QUEUE TO ACCESS THE ATM SERVICES.
         A POP-UP INTIMATES THE MOBILEAPP FIC HIS PRIORITY ORDER NUMBER.
         }
      ELSE
         {
         MOBILEAPP REMAINS IN PASSIVE EXECUTION MODE.
         }
      }
   ELSE
      {
      GPS_COORDINATES_MOBILE MATCHING WITH ATM_GPS_COORDINATES ACTION CONTINUES.
      }

END FUNCTION
```

Fig. 6. AI-SS Algorithm 1: C19PCL-ATM Mobile App Server.

```
FUNCTION (ATM_DOOR_OPEN_CLOSE)

PreCondition: FIC_MobileApp_Priority_Order_No based Pop-up message to enter in to the ATM.

    IF (FIC_FACE_DATA_ATMDOOR_CAMERA = FIC_FACE_DATA_FIDB)
            {
            SECOND_STEP_AUTHENTICATION CONFIRMED
            ATM_DOOR (OPEN)
            }
    ELSE
            {
            ATM_DOOR (CLOSE)
            }

END FUNCTION
```

Fig. 7. AI-SS Algorithm 2a: ATM door operations.

```
FUNCTION (DEPOSIT/WITHDRAWAL)

PreCondition: DISTANCE_FUNCTION_GPS (ATM and MOBILE = ONE METER)

IF (FIC_FI_ACCOUNT > 1)
    SELECT(FI_ACCOUNT)
ELSE
    SELECT(AVAILABLE_ACCOUNT)

//MOBILE_APP_SELECTION OPERATION

SELECTION = (WITHDRAWAL, DEPOSIT)
SWITCH (SELECTION)
        CASE 1: WITHDRAWAL
            {
            .............
            }
        CASE 2: DEPOSIT
            {
            .............
            }
        DEFAULT: GOTO_SELECTION
END SWITCH

END FUNCTION
```

Fig. 8. AI-SS Algorithm 2b: ATM operations.

The Mobile App server is equipped with the AI-SS2 algorithm. The C19PCL-ATM Mobile App Server algorithm is presented in Fig. 7.

The Mobile App server is equipped with the AI-SS algorithm 2b. The C19PCL-ATM Mobile App Server is presented in Fig. 8.

5. Future Enhancements

The scope of this work is wide open in computer science domains like data mining, data analysis, database management system, computer vision, and AI. As for the future enhancement of this work, the ATM's doorstep computer vision-based face recognition system provides many opportunities to explore image data analysis and data mining to optimize data retrieval and analysis techniques using new methodologies. The optimization-based algorithms can be applied in various parts of the proposed work to extend the scope of the research. The supervised learning-based machine learning algorithms can be further tested with the AI-SSs, to enhance the AI concept in the proposed work. In the future, the two-step authentication method used in the proposed work can be enhanced with the latest introduction of novel methods in terms of the global app, i.e., all available FIs can be merged with a single mobile app. The automatic door opening of C19PCL-ATM opens up a possibility for the mobile app service providers to integrate on a global scale by providing ATM services to all FIs through a single app. The scope of work in the proposed model was restricted to only two ATM operations. However, in the future, all available ATM operations could be implemented with the SD-based CL-ATM operations.

6. Conclusion

The SD is the primary factor of the COVID-19 pandemic situation, which prevents the infection/spread of the virus. The contact-based operations at the ATM are a potent source for the spread of COVID-19. The article, therefore, focused on providing a technological solutions to overcome such virus infections [16]. The article details the setup of the C19PCL-ATM and touch-free IAAPO operations. The practical implementation of the proposed design and further enhancements could enable all the necessary operations of the ATM.

References

[1] Ari Ramadhana, H. and Dieky, A. 2021. Modelling and verification of cash withdrawal transaction in automated teller machine using timed automata. pp. 1–10. *In*: International Conference on Mathematics: Pure, Applied and Computation, Journal of Physics: Conference Series, vol. 1821.

[2] Balaji, S., Vikas, H. and Pravin, Y. 2017. Dorsal hand vein authentication system: a review. Journal of Scientific Research and Development 6: 511–514.

[3] Karthika, S., Ambeth Kumar, V. D., Venkatesan, R., Sathyapreiya, V., Saranya, G. et al. 2019. Two three step authentication in ATM machine to transfer money and for voting application. Procedia Computer Science 165: 300–306.

[4] Mohsin, K., Saifali, K., Sharad, O. and Kalbande, D. R. 2015. Enhanced security for ATM machine with OTP and facial recognition features. Procedia Computer Science 45: 390–396.

[5] Kannan, G. 2017. Handheld secured electronic doorstep banking system that allows cash withdrawal and deposit facility for remote and rural areas. Indonesian Journal of Electrical Engineering and Computer Science 8(3): 705–708.

[6] Waqas, A., Fatima, N., Fatima, M., Siddiqi, S., Rehman, A. et al. 2019. ATM tracker: tracking the status of automatic teller machine for users' easiness. Proceedings of the 7th International Conference on Communications and Broadband Networking, pp. 67–71.

[7] Amit, K. and Varsha. 2019. ATM modeling through probabilistic and reliability approach subject to different failures. Journal of Engineering Science and Technology Review 12: 145–154.

[8] Imran, M. A., Mridha, M. F. and Nur, M. K. 2019. OTP based cardless transaction using ATM. International Conference on Robotics, Electrical and Signal Processing Techniques (ICREST), pp. 511–516.

[9] Olayeye Ilemabayo, O., Otuonye Anthony, I. and Eke Florence, U. 2020. Development of electronic bank deposit and withdrawal system using quick response code technology. International Journal of Scientific and Engineering Research 11(12): 345–356.

[10] Tewodros, A. and Debela, B. 2019. Factors affecting customers satisfaction towards the use of automated teller machines (ATMS): A case in commercial bank of Ethiopia, Chiro town. African Journal of Business Management 13: 438–448.

[11] Beata, Ś., Paweł, T. and Dominik, P. 2021. Transaction factors' influence on the choice of payment by Polish consumers. Journal of Retailing and Consumer Services 58.

[12] Pavel, L., Artyom, S., Elizaveta, L., Michael, E., Ekaterina, N. et al. 2020. The use of artificial intelligence technology in the process of creating an ATM service model. Procedia Computer Science 169: 203–208.

[13] Kholoud, Y., Mohammed Najmi, A., AlZain Masud, M., Jhanjhi, N. Z., Al-Amri, J. et al. 2021. A survey on security threats and countermeasures in IoT to achieve users confidentiality and reliability. Materials Today: Proceedings.

[14] Ankur, J. and Sushant, R. 2020. Implications of emerging technologies on the future of work. IIMB Management Review 32(4): 448–454.

[15] Adewale, S., Jacob, M., Josephine, M. and Suleiman, N. 2018. A simulation model for cardless automated teller machine transactions. i-manager's Journal on Digital Signal Processing 6(1).

[16] Iyabo, A. and Simon-Oke. 2019. Bacteriological evaluation of some automated teller machine. *In*: Akure Metropolis. Pan African Journal of Life Sciences 3: 183–187.

[17] Jillet, S., Anwesha, C. and Janaki, S. 2021. Cashlessness in India: Vision, policy and practices. Telecommunications Policy 45.

[18] Benyam, T. and Fayera, B. 2021. Effects of automated teller machine service on client satisfaction in Commercial Bank of Ethiopia. Heliyon 7(3): e06405.

[19] Balamurugan, C. R., Ramash Kumar, K. and Thirumalai, A. 2019. Secured smart ATM transaction. International Journal of Reconfigurable and Embedded Systems 8(1): 61–74.

[20] Yunhui, Z., Ziyi, Z., Hongfei, G., Yilin, C., Shiyue, S. et al. 2019. ATM transaction status feature analysis and anomaly detection. Studies in Engineering and Technology 6(1).

[21] Arun, S., Jhilik, B. and Shatrughan, M. 2017. Face as bio-metric password for secure ATM transactions. Proceedings of Sixth International Conference on Soft Computing for Problem Solving, Advances in Intelligent Systems and Computing, vol. 547.

[22] Chithra, P. L. and Ilakkiya, A. B. 2019. Bio-metric face detection techniques using HCA for ATM transactions. International Journal of Information Technology 3: 71–76.

[23] Abiodun, E. O., Aman, J., Oludare, I. A., Humaira, A., Nachaat, A. M. et al. 2019. Fingereye: improvising security and optimizing ATM transaction time based on iris-scan authentication. International Journal of Electrical and Computer Engineering 9: 1879–1886.

[24] Tay, B. and Mourad, A. 2020. Intelligent performance-aware adaptation of control policies for optimizing banking teller process using machine learning. IEEE Access 8: 153403–153412.

[25] Anoud, B. H., Munir, M. and Aisha, A. 2019. Online authentication methods used in banks and attacks against these methods. Procedia Computer Science 151: 1052–1059.

[26] Ameh, I. A., Olayemi, M. O. and Olumide, S. A. 2016. Securing cardless automated teller machine transactions using bimodal authentication system. Journal of Applied Security Research 11(4): 469–488.

[27] Afia, F., Noor, M,, Sharif, A., Jannat, M., Juena, N. et al. 2021. Trifecta approach to ATM transaction security. International Journal of Computer Applications 183(4): 18–23.

[28] Aleksander, B., Igli, T., Olimpjon, S., Elson, A., Alban, R. et al. 2021. Security of automated teller machines (ATM's). New Technologies, Development and Application IV. NT, Lecture Notes in Networks and Systems 233.

CHAPTER 11
Scope of Optimization in Plant Leaf Disease Detection using Deep Learning and Swarm Intelligence

*Vishakha A Metre** and *Sudhir D Sawarkar*

1. Introduction

It has been found that there are more than 1.7 million living species on the earth's surface; off these, the plant is regarded as one of the important natural entities which extends to capture a noteworthy portion of the earth; not only this but also is responsible for supporting human existence. With the use of modern technologies, humans can produce enough food to feed the entire world. Although enough food is produced, its variety is limited by several parasitic and non-parasitic causes. Diseases found in plants can be blamed for the shortages we are facing in terms of food worldwide, posing a threat to small farmers who rely on healthy crops yields to survive. Furthermore, worldwide millions of people have insufficient food. Besides, if we talk about the statistics of India, 17.5% of the harvest is lost annually to pests and diseases [1, 6, 12, 21].

Small farmers are crucial to agricultural output, and 80% of total crop yield comes from the population, but small farmers have lost 50% due to parasitic and non-parasitic diseases found in plants. Previously, the task of identifying the illnesses in crops, herbs, shrubs, and others, was done by plant clinics

Department of Computer Engineering, Datta Meghe College of Engineering, Airoli, Navi Mumbai, India.
Email: sudhir_sawarkar@yahoo.com
* Corresponding author: vishakha.metre@gmail.com

and agricultural organisations or institutions [6, 24, 25]. Visual inspection is the most common method used by specialists to identify plant diseases. Expert visual disease diagnosis necessitates frequent inspections, which most farmers cannot afford. Experts must also communicate with one another in some developing countries, and farmers must travel considerable distances. This prolongs and increases the cost of the consultation process. Farmers, for the most part, are uninformed of non-indigenous diseases, which forces them to use non-indigenous pesticides by consulting experts. Advances in object identification provides possibilities for smart phones and online applications; also provide farmers with robotics knowledge; develop more powerful and accurate disease detection applications [9, 30]. In object recognition, picture classification, natural language processing, and healthcare, recent advances in machine learning and deep neural networks are yielding great results and outperforming prior cutting-edge technology [2, 5, 29]. This study examines a deep learning (DL)-based approach to plant leaf disease identification, which overcomes the limitations of existing image processing methods while also allows for optimization and introduces a few experimental setups.

2. Plant Leaf Disease Detection

Plant disease detection is critical in agriculture, as farmers must frequently determine if the crop they are harvesting is fit for consumption. These should be addressed carefully because they might cause major difficulties in the systems, affecting the crop quality, quantity, or productivity. Plant diseases generate disease outbreaks on a regular basis, resulting in large-scale fatalities and a negative economic impact [22, 23]. To save people's lives and money, these issues must be addressed early on. In the realm of computer vision, detecting plant diseases is a critical component of study. This is regarded as a tool that evaluates the presence of any disease in the collected plant with the help of acquired photos of the target plant leaf. Currently, devices for the mechanical detection of plant diseases are initially used in agriculture and have to some extent replaced the traditional identification with the naked eye. Spontaneous organization of plant illnesses is an imperative study; since, taking intensive care of big harvests and detecting disease symptoms as soon as they develop is one of the critical tasks in the said area [25, 31]. This facilitates photographic inspections to be performed instantly using computational methods. Visual scanning by experts, on the other hand, is time-consuming, inaccurate, and can only be done in small regions at a time. Plant illnesses can be detected early with the new age technology [6, 15, 21].

2.1 Plant Leaf Disease Detection using Image Processing

One of the major challenges is identifying plant diseases from photographs of plant leaves in image processing and object recognition. The standard procedure for recognizing and classifying plant leaf images into healthy and unhealthy categories with disease names involves five important phases [9, 10, 11, 12]. The first is the image acquisition phase, where the images of the leaf are captured by any digital camera. Each captured image must be pre-processed with a visible transformation in order to have noise free images. Image segmentation is the third stage, which recovers the area of interest from the original image's background. The segmented image is used as an input to the fourth step, which contains crucial properties termed feature vectors, which are then fed to the classifier in the fifth stage. Almost all existing approaches to the aforementioned problem have gone through the same procedure in identifying plant illnesses, however their idea of improvement includes the employment of various image processing and machine learning algorithms at various phases of the process [7, 8, 21]. The Fig. 1 depicts the idea of the work.

The traditional approach to plant disease detection has used various image processing algorithms followed by the construction of feature vectors to help the classifier classify images into healthy and unhealthy. These approaches depended on the characteristics of the plant diseases, the appropriate light source, and the viewing angle. Although these approaches overcome the limitations of traditional algorithmic architecture, they incur implementation costs [13, 20, 34]. In addition, images collected in natural light also present several challenges; namely, the presence of noise in the image, minimum disparity, a wide range in the size of the lesion, small differences between the lesion area and the background. For these reasons, conventional methods often fail to produce better results. Machine learning approaches for the detection and classification of plant diseases take account of conventional computer visualisation techniques such as histogram of gradients, image

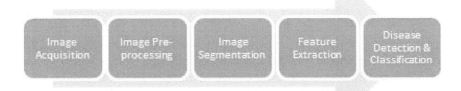

Fig. 1. Image processing based plant leaf disease detection.

grouping algorithms, SVM, KNN, K-means, and ANN in particular at various stages of the process [2, 5, 23, 26].

2.2 Plant Leaf Disease Detection using Deep Learning

In the recent decade, a lot of work has been done on using deep learning and computer vision to detect plant diseases. DL has received more attention due to its superiority in terms of accuracy when training large amounts of data. DL has prominently shown its application in various areas, including healthcare, computer vision, image classification, facial recognition, and many more. Today, many domestic and foreign companies are working on models based on deep learning for the effective detection of plant leaf diseases. Therefore, deep learning-based plant leaf disease detection has not only become an important academic research area, but also has a future from the point of view of application to the market [3, 4].

Deep neural learning (DNL) has lately been effectively applied in a variety of sectors for end-to-end learning. Neural networks create a link between an input (a picture of a diseased plant) and an output (a pair of plant diseases). The nodes of a neural network are computational models which are mathematical in nature that receive a numeric input from the leading edge and output a numeric output on the trailing edge. Deep neural networks simply use a succession of stacked nodal layers that takes the input layer and links it to the successive output layer. The goal is to build a deep network in which the network's structure, mathematical nodes along with the weights assigned to each link between the mathematical nodes, accurately links the input to its appropriate output [3, 27].

Deep learning is useful for recognising images with better precision and robustness because it does not require the extraction of specific features; nonetheless, the relevant qualities can be found through iterative training and learning. Conceptually, deep learning means using neural networks to analyse large amounts of data and show functional learning. Several hidden layers play a role in extracting all data features from the input image, and each hidden layer represents a perceptron that, compared to conventional image recognition methods, is responsible for extracting lower-level features that tare then combined to form higher-level characteristics. Traditional plant diseases rely on luck and experience, and cannot acquire the knowledge and recover features from the target image automatically. An opposite to the conventional methods, with the zero manual involvement; deep learning has an ability to start learning from enormous volumes of data. Furthermore, with a higher number of sample images for the process of training the model, the performance of deep neural learning improves remarkably [4].

Several deep learning approaches are currently available, including Convolutional Neural Networks (CNN), Deep Belief Networks (DBN), Sack De-noising Autoencoder (SDAE), and Deep Boltzmann Machine (DBM), among which the deep CNN is the most prevalent in deep learning frameworks. AlexNet, InceptionV3/GoogleNet, ResNet50, VGGNet, SqueezeNet, MobileNet, and other CNN models are used in Deep Learning-based plant disease classification models [1, 2, 5].

The method based on deep learning for the detection and classification of plant leaf diseases comprises three steps; Image acquisition to collect diseased plant leaf images, image editing or enlargement, application of various pre-processing/enlargement functions to rotate the images, cropping, filtering and more, to make a new set of enlarged images which can be optional but one good choice from an insufficient image dataset, followed by a feature extraction and classification stage that not only extracts all feature sets from the image but also uses them for classification purposes leading to the detection of diseases in the form of some plant diseases. Figure 2 shows the approach to the detection and classification of plant leaf diseases based on deep learning.

Although, a lot of debate and research is going on in the two above mentioned approaches for solving the problem of plant leaf disease detection, it is observed and experienced that the deep learning approaches are winning with a higher proportion of accuracy with less complexity. Figure 3 Summarizes the % accuracy graph of image processing (machine learning) and deep learning approaches studied in the research review and Table 1 states the differentiating factors between the two mentioned approaches.

Fig. 2. Deep Learning based plant leaf disease detection.

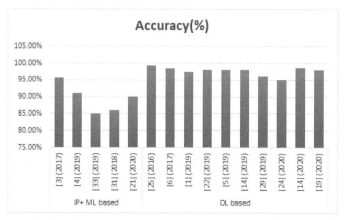

Fig. 3. Comparing accuracy of IP (ML) and DL based approaches for PLDD.

Table 1. Deep learning vs. image processing approach.

Deep Learning Approach	Parameters	Conventional Image Processing Approach
Applicable for huge amount of training dataset	Training Dataset size	Applicable for small size training dataset
Convolutional neural network (CNN), two types of approaches of CNN architecture building: 1. Training from scratch 2. Transfer Learning	Methodology	1. Image segmentation 2. Feature extraction method 3. Classification method:
Automatic learning of features from huge amount of data	Unique Feature	Manual design features + classifiers (or rules)
Sufficient amount of training data and computing units that provides high-performance.	Necessary condition	Prerequisite conditions for image processing are comparatively demanding, maximum disparity must be there between infectious part and healthy part of the leaf image, only minimal noise presence is tolerated.
Suitable for real time dataset and application since, change in environment, lighting effects, plant disease category won't bother the accurate detection and classification of plant diseases.	Application Scenarios	Changing threshold value as well as redesigning the algorithm is must if whenever environmental or plant disease category changes that results into poor results in case of real time dataset.
Can be applied for multi-class classification problems in terms of plant leaf disease detection.	Type of Problem	Can be applied for single class classification problem in terms of plant leaf disease detection.
No need of feature engineering at complex level.	Computing Complexity	Need of complex and accurate feature engineering.

3. Plant Leaf Disease Detection

Considering different optimal approaches for addressing multiclass plant leaf disease detection problems, a study of CNN architecture, concept of transfer learning (TL) approach and scope of using optimization methodology, i.e., particle swarm optimization (PSO) in CNN has been done from a research view point.

3.1 Convolutional Neural Network

In fact, the CNN model consists of 5 basic layers; first is input, second is convolutional followed by third layer, i.e., pooling, fourth is fully connected and fifth and final one is output. A convolution kernel is first defined in the convolution layer. During the phase of processing input data, the role of the convolution layer is to glide over the generated feature vector in order to retrieve some of the important details about the features. Neurons are pumped into the pooling layer after feature extraction from the convolutional layer in view of repeating the process of retrieving the features. One of the common methods of pooling today is the calculation of the average value, largest value and randomly chosen one among all the values. Once the data enters multiple convolutional layers followed by pooling layers, it enters the fully connected layer, and the neurons in the complete connection layer are totally connected to the neurons in the top layer. As a final point, the statistics in the complete connection layer are categorised using the well-known Softmax method, and categorized values are then transferred to the final output layer to publish the categorized results [2, 6]. Figure 4 shows the basic structure of the CNN model.

According to the different approaches accomplished with the help of the classification framework, its sub-division can be done into several categories which are mentioned in the Table 2 with their unique approach and pros and cons respectively.

In a true expected surrounding, the wide variations in contour, dimension, surface, color, pattern, and brightness of plant disease images categorizes the recognition task as the most tedious one. CNN, being the best feature extractor; has also achieved its name in the classification task of plant diseases.

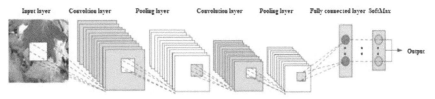

Fig. 4. Basic structure of CNN model [6].

Table 2. Different CNN classification network methods.

CNN Methods	Approach	Con's
CNN as Classifier for Original Images	Takes complete set of plant disease images as it is for the sake of training. It can acts as a basis for other classification architectures.	Infectious portion must make up a particular percentage of the image; else, their physical appearance will be easily merged and not suitable for multi class lesions.
Multi-category classification	If the classes of plant disease exceed 2, this is applicable. This approach is suitable for multi-objective plant diseases classification framework.	Need of secondary training is must.
CNN as Classifier after identifying Region of Interest (ROI)	Its main target is to check whether there is any infectious appearance in the image to identify the region of interest which can be then feed to the machine to predict the disease.	Additional headache is to employ appropriate methodology for extracting ROI.
Multi-task learning network	This framework not only carries the output generated from segmentation phase but also from classification phase that together called as multiple task training framework. Good choice for reducing number of training samples requires.	1. Complex architecture. 2. Need of pixel by-pixel label in adding branch wise segmentation.
CNN as Feature Extractor	It uses a pretrained model of CNN architecture using transfer learning concept in order recover features from the input images which are then given to ML classifier for classification purpose. Appropriate in finding effective features with respect to lesion.	Dependency on other classifiers for final classification results
Sliding window	A repetitive sliding approach over input images within a small size window is given to classifier. Lastly, all the generated sliding windows are merged together to find the infectious portion of the image.	1. Accurate selection of size of Sliding window size is must. 2. It can get rough position not actual. 3. Slow in traversal and sliding.
Heatmap	This constitutes an image that focuses on each and every important portion of the image by measuring the darkness, darker the color states more importance of that specific region, i.e., greater the probability that it is the lesion. More accurate in generating actual lesions areas.	Identifying location of accurate lesions is relying on performance of classification architecture.

3.2 Transfer Learning

A previously skilled model is a stored framework that has been previously learned from a huge data set, typically a comprehensive image grouping task. One can use the previously skilled model as it is or make use of the transfer learning concept to adapt it for a particular task. The intuition behind employing transfer learning based CNN architecture for the purpose of grouping images is that when a model is efficient enough after training with an enormous and sufficiently generalized data set, such a model stands as a generic model in this competitive computerized visualization world.

The key benefit of employing transfer learning is that it allows you to learn more quickly and at the same time such a model does not start the learning process from scratch, but starts from learned patterns in solving another problem; which is similar to the targeted one we are supposed to solve using previously trained models. Classification of images can be addressed as a notable problem which can be solved using such an approach of transferred knowledge. The said approach is useful in optimizing the process of training the machine with a dataset by employing an optimal methodology in the current approach. There are two ways to adapt a previously trained model [3, 4].

3.2.1 Feature Extraction

This is the process where relevant and useful information such as characteristic features of a new sample image fed as an input are acquired with the help of previously owned knowledge. This idea requires building an innovative classifier layer educated from the ground up in addition to the previously trained network architecture leading to the reuse of the previously learned feature maps for the dataset. This saves the process of complete training of the model architecture. Although, the basic convolution network has sufficient knowledge about how to classify the images, the newly added classifier is meant for the special purpose of classification of the targeted image classes. The concept of using deep learning architecture can be a better choice to extract features more efficiently and quickly, as well as with the help of the transfer learning approach, which uses pre-trained models from the ImageNet dataset, one of the largest image datasets that eliminates the headache of training which shortens the training time. There are numerous options for deep neural network architectures, but very few of them have been tested in the literature. Various models of deep neural networks such as AlexNet, VGG16/19, ResNet50, GoogleNet/Inception, MobielNet, SqueezNet, among others, can be experimented with for feature extraction purpose [1].

3.2.2 Fine-Tuning

Release few of the uppermost layers of a frozen model base and conjointly train not only the newly-added classifier layers but also the final layers of the basic model. This allows us to "fine-tune" the underlying model's higher-order feature representations to make them more relevant for the task at hand.

3.3 Optimization Methodology

Swarm intelligence (IS) emerged as an innovative branch of artificial intelligence (AI) that is based on observing the behaviour of various natural beings. The basic idea is to encourage social behaviour within ant colonies, flocks of birds, hives, schools of fish, and others, an exceptional solution to various complex problems. The different optimization paradigms motivated by swarm intelligence are listed in alphabetical order of their names as ant colony optimization (ACO), bee colony optimization (BCO), fish school optimization (FSO), particle swarm optimization (PSO), current gray wolf optimization (GWO), etc., are discovered in the SI taxonomy. Figure 5 shows the SI scenario. Although various optimization algorithms have been studied and used in many notable applications, PSO has its own relevance in handling complex and nonlinear optimization problems.

PSO, has attracted a lot of research attention today due to its simplicity and effectiveness of implementation. It is popularly known as the population-based stochastic search algorithm for solving swarm intelligence (SI) domain multi-peak computational problems. It is also a recognized evolutionary calculation algorithm that simulates the movement of birds in their flocks that perform a global search in the wide solution space. PSO is based on a very few parameters, leading to fast and accurate calculation results, demonstrating its importance in SI. Therefore, PSO can be described as an excellent option to solve problems with clustering, segmentation and classification approaches [7, 33]. PSO can be used primarily after the feature extraction process for optimal selection of feature subsets, as well as in the classification phase to

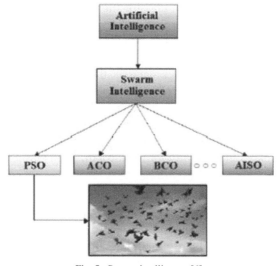

Fig. 5. Swarm intelligence [6].

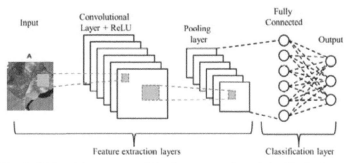

Fig. 6. CNN model structure specifying feature extraction and classification layers.

optimize hyper parameters and ensure higher precision. Figure 6 shows the structure of the CNN model, showing two phases in which optimization can be applied.

3.3.1 PSO in Feature Selection

The performance of an image classifier depends on its ability to extract a good number of features that carefully characterize the image. The feature extractor also plays an important role in plant leaf disease detection in order to train the classifier to classify the input image into the correct plant disease pair category [16, 17]. Once the features have been extracted from a deep neural network model, the classifier with the best features becomes more accurate if an optimal selection of features can be made using one of the optimization methods, while saving training time with limited feature training. The main purpose of selecting the characteristic features is to make an optimal choice of relevant features while upholding the correctness of the classification [1, 24, 31, 32]. The procedure to do this is as follows and its flow chart is shown in Fig. 7.

1) Use of deep neural network with transfer learning approach for feature extraction.

2) SoftMax layer of the model architecture is replaced by the desired number of features for the model.

3) The ultimate goal of PSO is to scan all the characteristics features but selecting the subset of the most expressive features.

4) Every single element in the swarm characterizes a probable candidate solution and the objective function to be optimized can be the fitness function which makes all the elements explore their options before it converges to best solution.

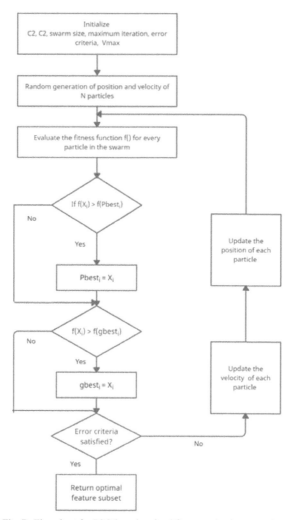

Fig. 7. Flowchart for PSO based optimal feature selection procedure.

5) The initial setting of necessary parameters is randomly produced and every particle in the swarm is represented as P = F1F2F3F4......Fn, where n = 1, 2, 3,..., m and m is the feature length vector.

6) Each index of the binary string B represents which feature is to be selected, 1 – selected and 0 – rejected.

  Eg. For n = 10 dimensional dataset

   P = F1F2F3F4F5F6F7F8F9F10

   PSO selected Feature Subset by setting bits 2, 5, 7, 8 and 9 to 1: F2F5F7F8F9

7) It is very important to note that the fitness function plays an important role in preserving the efficiency of the generated subset of the characteristic features and it also acts as a key measure to compare it with respect to the fitness value of the original set of acquired features in each iteration of the PSO.

8) In each and every iteration of the PSO, 1,2,---,m indices of the particle indicate the evolution of parameters and their quality assessment is done using the objective function value.

9) This assessment is done with respect to the target of achieving a promising value from the classification task.

3.3.2 PSO for Hyper Parameter Optimization of CNN Classifier

CNNs are mostly regarded as one of the best choices for visualization applications because of their highly promising results; but at the same time we cannot be ignore the fact that they invite huge computation costs [18]. This further invites the novel implementation to be researched in order to optimize it. CNN architectures consist of plenty constraints for which a proper configuration value has to be decided wisely; since if a change in the parameter values occurs they will differ from the intended results for the same problem. Values for hyper parameters are set on a per-hyper parameter basis; typically applying an arbitrary searching strategy, running multiple assessments, or manual tuning which is a complex process; requiring CNN parameter optimization to improve detection rate and reducing computational effort. To overcome the said obstacle, it is observed in the literature review that computational strategies which are evolutionary in nature are capable enough to inevitably project the optimal design of the CNN model and improve their performance [2, 19]. The numerous parameters in the individual CNN layers are shown in Fig. 8.

PSO can be used to improve the necessary parameters of the CNN architecture. To identify suitable parameters, first go for the most significant factors that control CNN performance, and then use the PSO method to find them [14, 15]. With the repetitive experiments; by manual setup, one can finalize the list of parameters that require attention in order to maintain a good model performance.

The below mentioned specify the details of the experimentation done with respect to the suitable hyper parameters for the proposed work:

- The count of convolutional layers: Let x1 with search space [1, 3]
- The size of filter per convolutional layer: Let x2 with search space [32, 128]

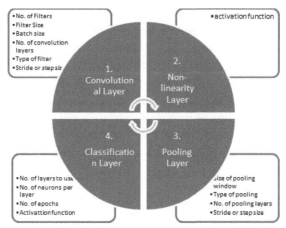

Fig. 8. Different hyper parameters in different layers of CNN.

- The count of filters: Let x3 with search space [1, 4]
  such as for value of x3 = 1, search space [3, 3],
        for value of x3 = 2, search space [5, 5],
        for value of x3 = 3, search space [7, 7],
        for value of x3 = 4, search space [9, 9],
- The batch size: Let x4 with search space [32, 256]

The general process of optimizing CNN with PSO is shown in Fig. 9. The "Training and Optimization" section is the one in which the CNN has been set up to incorporate parameter optimization using the PSO algorithm. The PSO is set up for the run using the parameters that have been supplied and particles are generated from it. In the PSO approach, every single particle is eligible to be called a probable solution and its associated location mentions the parameter to be optimized. Thus, one complete iteration of the CNN training process is represented by all probable solutions.

The flowchart of PSO-CNN approach is illustrated [2] in the Fig. 10 and its explanatory steps are given as below:

1) Input image dataset to train the CNN: Select the image data set for PSO-CNN procedure to be implemented. It is important to keep a similar colour gamut and scale; also dimensions and file format has to be same for all the images.

2) Generation of the PSO parameter population: It involves setting the static parameters for the PSO using trial and error, i.e., iteration count, particle count, parameter for inertia weight (W), self-confidence constant value (C1) and social confidence constant value (C2).

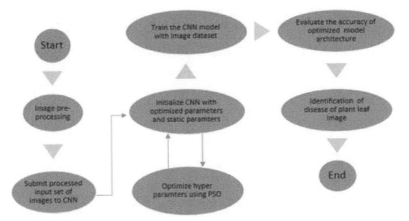

Fig. 9. General block diagram for PSO optimized CNN process.

3) Initialization of the CNN architecture: Get the CNN architecture ready with the parameters received from PSO, namely; count of convolution layers, the size of filters, the count of filters per convolution layer, and the size of batches along with the static parameters; knowledge earning function (e.g., Adam/SGD), classifier layer activation function (e.g., SoftMax), non-linearity activation function (e.g., ReLU), count of epochs (e.g., 5–20).

4) CNN training and validation: CNN categorizes the input image dataset into three parts; namely training image set, validation image set and testing image set. This in turn can act as an objective function for the PSO to attain intended classification accuracy.

5) Evaluate the objective function: PSO is meant to optimize the results by continuous evaluation of the considered objective function.

6) Update PSO parameters: With every iteration progress, update the speed and location value for each particle in the entire swarm considered; using the influence of the best position achieved by the particle with self-confidence (Pbest) and social confidence (Gbest).

7) Iterate till the stopping criteria: The entire process gets repeated for continuous evaluation of swarm particles until either the stop criteria is matched or the number of iterations are exhausted whichever is earlier.

8) Output optimal solution: Finally, Gbest specifying particle is returned as the ideal solution for the proposed CNN network.

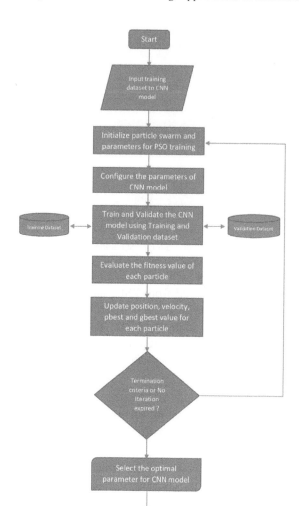

Fig. 10. Flowchart for PSO optimized CNN.

4. Experimental Study

4.1 Technical Specification

The most widely used deep learning tools are none other than Tensorflow along with Numpy, Pandas and Matplotlib, are utilized in the proposed work. Google Tensorflow also supports CPU, GPU, and mobile hardware. In addition, its features namely; portability, flexibility, performance among others, make us highlight it as the most widely used deep learning tool in C and Python application interfaces. For implementation purpose, both Jupiter notebook and Google colab have been explored.

4.2 Dataset

A publicly available dataset called PlantVillage is used, containing 54,305 images of diseased and healthy plant leaves collected under controlled conditions. This dataset consists of images of 14 plant species (apple, blueberry, cherry, corn, grape, orange, peach, bell pepper, potato, raspberry, soyabean, pumpkin, strawberry, and tomato), 38 class names specifying the pair of plant diseases, and 16 basic diseases, 4 bacterial diseases, 2 mould diseases, 2 viral diseases and 1 mite disease. Namely, the various diseases covered in this data set are; Scabies, rust, powdery mildew, early blight, late blight, black rot, measles, bacterial blight, mould and leaf curl, are associated with various types of plants. For the purpose of implementing this work a total of 8236 images of different plant diseases covering 38 class names was used for training due to computational limitations. Figure 11 shows one sample image from all 38 class labels [1, 3].

Fig. 11. Sample images of 38 class labels from PlantVillage dataset [3].

4.3 Proposed Architecture

The proposed architecture, which is a supervised machine learning project, begins with data collection for training purposes. The training dataset includes images of healthy and diseased plant leaves of various crops and the diseases associated with them. The next step is to clean and pre-process the image data as a prerequisite using the tf dataset and data augmentation methods. The data expansion process is essential to rotate, mirror and contrast the images in case we do not have enough images to train. We can also use Tensorflow's ImageDataGenerator method for the same purpose more efficiently. Once this is done, we can continue with the construction of the model and, for this, we intend to use a PSO optimized CNN or a related model optimized by PSO using transfer learning, since it is more suitable for the classification of images as mentioned above. We can then export the trained model to our hard drive. Once the optimized model is ready with efficiency, accuracy in minimized training time we can deploy it on a website or mobile application through the google cloud platform (GCP). The proposed approach is presented in the Fig. 12.

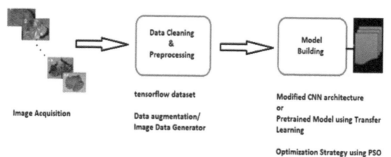

Fig. 12. Proposed architecture for PSO optimized plant leaf disease detection.

4.4 Results

The results of optimized CNN architecture are presented next and depicted in Fig. 13(a) and its relevant training and validation accuracy as well as loss is shown in the graph with Fig. 13(b).

CNN Model Version-1:

Input:

```
IMAGE_SIZE = 256, BATCH_SIZE = 32, CHANNELS = 3,
EPOCHS = 50
Training Dataset: 2150 images belonging to three
classes;
Class Labels: ['Potato___Early_blight', 'Potato___
Late_blight', 'Potato___healthy']
Train, Val, Test Split: 80%, 10%, 10%
Model: "sequential"
```

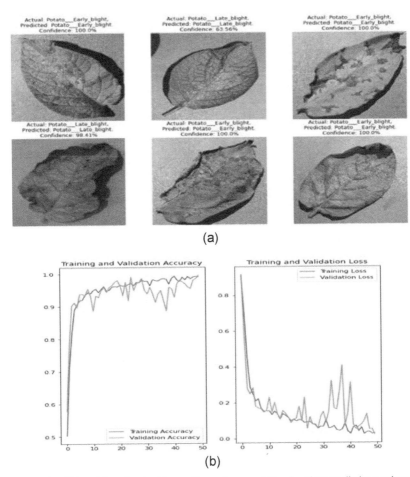

(a)

(b)

Fig. 13. (a) CNN model version 1 with rescaling and augmentation functions prediction results with actual class, predicted class and confidence for sample potato plant disease images. (b) CNN model version 1 with rescaling and augmentation functions training and validation accuracy and loss after 50 Epochs for sample potato plant disease.

Layer (type)	Output Shape	Param #
conv2d (Conv2D)	(None, 254, 254, 32)	896
max_pooling2d (MaxPooling2D)	(None, 127, 127, 32)	0
conv2d_1 (Conv2D)	(None, 125, 125, 64)	18496
max_pooling2d_1 (MaxPooling2)	(None, 62, 62, 64)	0

```
conv2d_2 (Conv2D)                    (None, 60, 60, 64)    36928

max_pooling2d_2 (MaxPooling2) (None, 30, 30, 64)    0

conv2d_3 (Conv2D)                    (None, 28, 28, 64)    36928

max_pooling2d_3 (MaxPooling2) (None, 14, 14, 64)    0

conv2d_4 (Conv2D)                    (None, 12, 12, 64)    36928

max_pooling2d_4 (MaxPooling2) (None, 6, 6, 64)      0

conv2d_5 (Conv2D)                    (None, 4, 4, 64)      36928

max_pooling2d_5 (MaxPooling2) (None, 2, 2, 64)      0

flatten (Flatten)                    (None, 256)           0

dense (Dense)                        (None, 64)            16448

dense_1 (Dense)                      (None, 3)             195
=================================================================
Total params: 183,747
Trainable params: 183,747
Non-trainable params: 0
```

Output:
```
loss: 2.92%
accuracy: 98.96%
val_loss: 2.19%
val_accuracy: 99.48%
```

The results of the optimized transfer learning approach using VGG16 architecture is presented next and depicted in Fig. 14(a) and its relevant training and validation accuracy as well as loss is shown in the graph with Fig. 14(b).

CNN Model (Transfer Learning VGG16) Version-2:

Input:
```
IMAGE_SIZE = 256, BATCH_SIZE = 32, CHANNELS = 3,
EPOCHS = 20
```

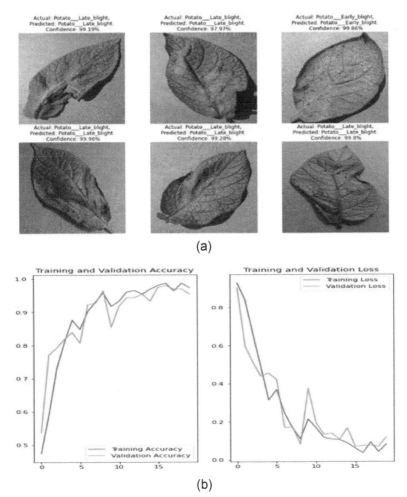

Fig. 14. (a) CNN model version 2 with transfer learning VGG16 architecture functions prediction results with actual class, predicted class and confidence for sample potato plant disease images. (b) CNN model version 2 with transfer learning VGG16 architecture functions training and validation accuracy and loss after 20 Epochs for sample potato plant disease images.

```
Training Dataset: 1506 images belonging to three
classes;
Class Labels: ['Potato___Early_blight', 'Potato___
Late_blight', 'Potato___healthy']
Train, Val, Test Split: 70%, 20%, 10%
Model: "model_2"
```

Layer (type)	Output Shape	Param #
input_2 (InputLayer)	[(None, 256, 256, 3)]	0
block1_conv1 (Conv2D)	(None, 256, 256, 64)	1792
block1_conv2 (Conv2D)	(None, 256, 256, 64)	36928
block1_pool (MaxPooling2D)	(None, 128, 128, 64)	0
block2_conv1 (Conv2D)	(None, 128, 128, 128)	73856
block2_conv2 (Conv2D)	(None, 128, 128, 128)	147584
block2_pool (MaxPooling2D)	(None, 64, 64, 128)	0
block3_conv1 (Conv2D)	(None, 64, 64, 256)	295168
block3_conv2 (Conv2D)	(None, 64, 64, 256)	590080
block3_conv3 (Conv2D)	(None, 64, 64, 256)	590080
block3_conv4 (Conv2D)	(None, 64, 64, 256)	590080
block3_pool (MaxPooling2D)	(None, 32, 32, 256)	0
block4_conv1 (Conv2D)	(None, 32, 32, 512)	1180160
block4_conv2 (Conv2D)	(None, 32, 32, 512)	2359808
block4_conv3 (Conv2D)	(None, 32, 32, 512)	2359808
block4_conv4 (Conv2D)	(None, 32, 32, 512)	2359808
block4_pool (MaxPooling2D)	(None, 16, 16, 512)	0
block5_conv1 (Conv2D)	(None, 16, 16, 512)	2359808
block5_conv2 (Conv2D)	(None, 16, 16, 512)	2359808
block5_conv3 (Conv2D)	(None, 16, 16, 512)	2359808
block5_conv4 (Conv2D)	(None, 16, 16, 512)	2359808

block5_pool (MaxPooling2D)	(None, 8, 8, 512)	0
flatten_1 (Flatten)	(None, 32768)	0
dense_1 (Dense)	(None, 39)	1277991

```
Total params: 21,302,375
Trainable params: 1,277,991
Non-trainable params: 20,024,384
```

Output:

```
loss: 8.12%
accuracy: 97.26%
val_loss: 1.11%
val_accuracy: 95.31%
```

5. Conclusion

Plant diseases are a major threat to the food supply around the world. Saving crops from early diseases will help farmers take preventive and control measures. As technology advances, an automated framework for detecting plant leaf diseases can be implemented to help farmers. Deep learning has received increasing attention due to its superiority in terms of accuracy when training large amounts of data. It has its own meaning for image recognition with higher precision and high robustness, since it is not necessary to extract certain characteristics; however, through iterative training and learning, you are able to find the right characteristics. Compared to old computer vision methods, which are made up of multiple steps and combinations, in order to achieve the disease detection task in the input plant leaf image; deep learning methods on the other side, for plant disease detection combine them in a start-to-finish approach for extracting features that has extensive development perception. Therefore, the study was conducted to examine the concepts of convolutional neural networks, transfer learning, and the possibilities of optimizing deep neural networks.

Two approaches have been proposed for optimization in deep neural networks, the first to use the particle swarm optimization algorithm in the feature selection process to obtain a reduced subset of features in order to further train the classifier for better accuracy. Second, it's about optimizing several important hyper parameters; namely, the count of layers for the convolution process, the size of the filter per convolutional layer, the count of filters used for convolution, and the size of batches employed using the particle swarm optimization algorithm. The proposed architectural block diagram is presented and experiments are performed to demonstrate the data acquisition and model building phases of the proposed architecture. Both the CNN model and one of the pre-trained models (VGG16) with transfer learning were experimented with the PlantVillage dataset with a limited number of images of healthy and diseased leaves and achieved an accuracy of 98.96% and 97.26%, respectively. Although, model-1 is superior in terms of accuracy and loss, it took more time as compared to model 2 which has avoided training from scratch due to the use of transfer learning. For future work, more relevant models will be experimented with an expanded data set along with the construction of a new CNN model with implementation of the optimization. Also, future work will be focusing on reducing the training time with an optimal approach.

References

[1] Yadav, Rishabh, Yogesh Kumar Rana and Sushama Nagpal. 2018. Plant Leaf Disease Detection and Classification Using Particle Swarm Optimization. pp. 294–306. In International Conference on Machine Learning for Networking, Springer, Cham.

[2] Fregoso, Jonathan, Claudia I. Gonzalez, Gabriela E. Martinez et al. 2021. Optimization of convolutional neural networks architectures using PSO for sign language recognition. Axioms 10(3): 139.

[3] Sagar, Abhinav and Dheeba, J. 2020. On using transfer learning for plant disease detection. bioRxiv.

[4] Mohanty, Sharada P., David P. Hughes and Marcel Salathé. 2016. Using deep learning for image-based plant disease detection. Frontiers in Plant Science 7: 1419.

[5] Qolomany, Basheer, Majdi Maabreh, Ala Al-Fuqaha, Ajay Gupta, Driss Benhaddou et al. 2017. Parameters optimization of deep learning models using particle swarm optimization. pp. 1285–1290. In 2017 13th International Wireless Communications and Mobile Computing Conference (IWCMC), IEEE.

[6] Liu, Jun and Xuewei Wang. 2021. Plant diseases and pests detection based on deep learning: a review. Plant Methods 17(1): 1–18.

[7] Metre, Vishakha A. 2021. Research review on plant leaf disease detection utilizing swarm intelligence. Turkish Journal of Computer and Mathematics Education (TURCOMAT) 12(10): 177–185.

[8] Sheelavantamath, Bindu. 2020. Plant disease detection and its solution using image classification. International Journal of Futures Research and Development 1(1): 73–79.

[9] Singh, Vijai and Ak K. Misra. 2017. Detection of plant leaf diseases using image segmentation and soft computing techniques. Information Processing in Agriculture 4(1): 41–49.

[10] Pravin Kumar, S. K., Sumithra, M. G. and Saranya, N. 2021. Particle Swarm Optimization (PSO) with fuzzy c means (PSO-FCM)–based segmentation and machine learning classifier for leaf diseases prediction. Concurrency and Computation: Practice and Experience 33(3): e5312.

[11] Sethy, Prabira Kumar, Nalini Kanta Barpanda and Amiya Kumar Rath. 2019. Detection and identification of rice leaf diseases using multiclass SVM and particle swarm optimization technique. Int. J. Innovative Tech. and Exploring Eng. (IJITEE) 8(6S2): 108–120.

[12] Kaur, Prabhjeet, Sanjay Singla and Sukhdeep Singh. 2017. Detection and classification of leaf diseases using integrated approach of support vector machine and particle swarm optimization. International Journal of Advanced and Applied Sciences 4(8): 79–83.

[13] Singh, Vijai. 2019. Sunflower leaf diseases detection using image segmentation based on particle swarm optimization. Artificial Intelligence in Agriculture 3: 62–68.

[14] Chanda, Moumita and Mantosh Biswas. 2019. Plant disease identification and classification using back-propagation neural network with particle swarm optimization. pp. 1029–1036. In 2019 3rd International Conference on Trends in Electronics and Informatics (ICOEI), IEEE.

[15] Aravinda, C. V. 2017. Classification and clustering of infected leaf plant using k-means algorithm. pp. 468–474. In International Conference on Cognitive Computing and Information Processing, Springer, Singapore.

[16] Suja Radha. 2017. Leaf disease detection using image processing. Journal of Chemical and Pharmaceutical Sciences 10(1): pp. 670–672. ISSN: 0974-2115.

[17] Maity, Subhajit, Sujan Sarkar, Vinaba Tapadar, A., Ayan Dutta et al. 2018. Fault area detection in leaf diseases using k-means clustering. pp. 1538–1542. In 2018 2nd International Conference on Trends in Electronics and Informatics (ICOEI), IEEE.

[18] Hughes, David and Marcel Salathé. 2015. An open access repository of images on plant health to enable the development of mobile disease diagnostics. arXiv preprint arXiv:1511.08060.

[19] Senthilkumar Meyyappan and Sridhathan Chandramouleeswaran. 2018. Plant Infection Detection Using Image Processing. Article July 2018.

[20] Jasim, Marwan Adnan and Jamal Mustafa AL-Tuwaijari. 2020. Plant leaf diseases detection and classification using image processing and deep learning techniques. pp. 259–265. In 2020 International Conference on Computer Science and Software Engineering (CSASE), IEEE.

[21] Padmanabhuni, Ms Sri Silpa and Pradeepini Gera. 2020. An extensive study on classification based plant disease detection system. Journal of Mechanics of Continua and Mathematical Sciences 15(5).

[22] Khairnar, Khushal and Nitin Goje. 2020. Image processing based approach for diseases detection and diagnosis on cotton plant leaf. In Techno-Societal 2018, pp. 55–65. Springer, Cham.

[23] Chanda, Moumita. 2019. Plant diseases classification using feature reduction, BPNN and PSO. International Journal of Software Engineering and Computer Systems 5(2): 90–103.

[24] Joshi, Barkha M. and Hetal Bhavsar. 2020. Plant leaf disease detection and control: A survey. Journal of Information and Optimization Sciences 41(2): 475–487.

[25] Ashish, Patil and Patil Tanuja. 2017. Survey on detection and classification of plant leaf disease in agriculture environment. International Advanced Research Journal in Science, Engineering and Technology 4(4).

[26] Kaur, Sukhvir, Shreelekha Pandey and Shivani Goel. 2019. Plants disease identification and classification through leaf images: A survey Archives of Computational Methods in Engineering 26(2): 507–530.

[27] Arsenovic, Marko, Mirjana Karanovic, Srdjan Sladojevic, Andras Anderla et al. 2019. Solving current limitations of deep learning based approaches for plant disease detection. Symmetry 11(7): 939.

[28] Sethy, Prabira Kumar, Nalini Kanta Barpanda and Amiya Kumar Rath. 2019. Detection and identification of rice leaf diseases using multiclass SVM and particle swarm optimization technique. Int. J. Innovative Tech. and Exploring Eng. (IJITEE) 8(6S2): 108–120.

[29] Kanaka Durga, N. and Anuradha, G. 2019. Plant disease identification using SVM and ANN algorithms. International Journal of Recent Technology and Engineering (IJRTE), ISSN: 2277-3878, Volume-7, Issue-5S4, February 2019.

[30] Kumar, Arun, Vinod Patidar, Deepak Khazanchi and Poonam Saini. 2016. Optimizing feature selection using particle swarm optimization and utilizing ventral sides of leaves for plant leaf classification. Procedia Computer Science 89: 324–332.

[31] Muthukannan, Kanthan and Pitchai Latha. 2015. A PSO model for disease pattern detection on leaf surfaces. Image Analysis & Stereology 34(3): 209–216.

[32] Sonal P. Patel and Arun Kumar Dewangan. 2017. A Comparative Study on Various Plant Leaf Diseases Detection and Classification. International Journal of Scientific Research Engineering & Technology (IJSRET), ISSN 2278–0882, Volume 6, Issue 3, March 2017.

[33] Vagisha Sharma, Amardeep Verma and Neelam Goel. 2020. Classification techniques for plant disease detection. International Journal of Recent Technology and Engineering (IJRTE) ISSN: 2277-3878, Volume-8 Issue-6, March 2020.

[34] Revathi, P. and Hemalatha, M. 2014. Identification of cotton diseases based on cross information gain deep forward neural network classifier with PSO feature selection. International Journal of Engineering and Technology 5(6): 4637–4642.

CHAPTER 12

Automatic Speech Analysis of Conversations for Dementia Detection Using Bi-LSTM Model

Neha Shivhare,[1,*] *Aditi Rai,*[2] *Shanti Rathod*[3] and *M R Khan*[4]

1. Introduction

DEMENTIA is a classification of neurodegenerative sicknesses that involves a long haul and generally steady lessening of intellectual working. It is described by a bunch of side effects that incorporate cognitive decline, thought challenges, faulty chief capacities (for example critical thinking, dynamic, arranging), language disability, engine issues, absence of inspiration and passionate misery. All through the sickness, the seriousness of these manifestations increments to the detriment of the patient's independence, just as their prosperity and their guardians' [1]. Monetary help for this exploration comes from the European Union's Horizon 2020 examination and advancement program, under the award arrangement No. 769661, towards the SAAM project; and from the Medical Research Council (MRC; award No. MR/N013166/1). Original copy got April 19, 2005; amended August 26, 2015, outcome of the neuropathology of various sicknesses, like Alzheimer's Disease (AD; half of dementia cases), cerebrovascular infection (25% of cases, including those that additionally show AD), Lewy body illness (15%

---
[1] PhD Research Scholor, Dr. C V Raman University, Kota, Bilaspur C. G., India.
[2] First Year M.Tech Student (Data Science and Analytics), IIIT, Allahabad (U.P.).
[3] Associate Professor, Electronics and Telecommunication department, Dr. C V Raman University, Kota, Bilaspur C.G. India.
[4] Professor, Electronics and Communication department, GEC Jagdalpur C.G. India.
Emails: ids2021902@iiita.ac.in; rathoreshanti@gmail.com; mrkhan@gecjdp.ac.in
* Corresponding author: neharai10@gmail.com, neharshivhare@gmail.com

cases), and other mind infections (5%), including Parkinson's infection, frontotemporal dementia and stroke [2].

The principal hazard factor for dementia is age, and in this manner its most prominent occurrence is among the older. Since the populace more than 65 years of age is anticipated to significantly increase between years 2000 and 2050 [3], dementia care is projected to have a gigantic cultural effect. In 2015, the WHO [4] assessed around 47.5 million instances of dementia around the world, with longitudinal accomplice concentrates on tracking down a yearly frequency somewhere in the range of 10 and 15 cases for each 1,000 individuals, from which somewhere in the range of 5 and 8 would be brought about by Alzheimer's Disease. The forecast is troublesome, with around 7 years of normal daily routine anticipation and under 3% patients experiencing longer than 14 years after conclusion [4]. Because of the seriousness of the circumstance around the world, foundations and specialists are contributing extensively to dementia anticipation and early identification, zeroing in on illness movement. There is a requirement for practical and adaptable strategies for the location of dementia from its most unobtrusive structures, like the preclinical phase of Subjective Memory Loss (SML), to more extreme conditions like Mild Cognitive Impairment (MCI) and Alzheimer's Dementia (AD) itself. The neuropathology of AD comprises of a few marvels, including intracellular aggregation of tau-proteine strands [5] and extracellular collection of beta-amieloid plaques [6]. Both are liable for cerebrum harm and neural utilitarian disturbance [7]. Such neuropathologhy is known to begin quietly as long as 20 years before a singular shows self-evident and recognizable intellectual indications, and there is no agreeable treatment for them. Hence, it is vital to discover procedures to recognize the issue as right on time as could really be expected, to improve treatment adequacy and personal satisfaction [8]. This review centers around AD acknowledgment utilizing acoustic data separated from unconstrained discourse. While cognitive decline is habitually viewed as the most unmistakable side effect of AD [9], discourse and language adjustments are additionally normal [10, 11]. Patients with AD as a rule show naming and word-discovering challenges (anomia) prompting aversion. Writing additionally recommends that patients with AD experience issues getting to semantic data purposefully, prompting an overall semantic weakening [12].

The heterogeneity of the indicative articulation of AD requires determination strategies that can catch more subtler viewpoints than ordinary screening instruments, which frequently neglect to segregate these side effects in pre-clinical AD. Social sign handling advancements are setting out open doors for individual wellbeing checking and improvement of symptomatic help instruments dependent on robotized preparing of conducting signals [13]. Discourse and language are rich and omnipresent wellsprings of intellectual social information, where computational investigation can possibly help clinicians ahead of schedule and precise determination of dementia [14]. There are a few normally utilized intellectual evaluations for dementia analysis that include etymological tests like the Mini-Mental State Examination (MMSE) [15], the fiveword test [16], the front facing appraisal battery [17], and the instrumental exercises of every day living scale [18]. Discourse coherence, for example, might be evaluated through picture depiction errands [19] or through commencement assignments [20], and Semantic Verbal Fluency (SVF) normally includes naming undertakings [21]. Nonetheless, while still important for determination, the majority of these neuropsychological tests offer little knowledge into the beginning phases of neurodegeneration and henceforth there is an expanding interest in creating elective strategies for early recognition. For example, the concentrate by Konig et al. [20], recorded a sentence rehashing task and utilized unique time cautioning to assess the waveforms. They took a gander at the arrangement bend between sets of relating waveforms to see whether there is a huge contrast between the sentences created by the clinician (apparently sound) and the sentences delivered by the AD patients [20]. A weakness of these tests is that they utilize discourse and language created under controlled research facility conditions instead of unconstrained discourse, which would be needed for pragmatic longitudinal screening and checking in daily life. One of a handful of unconstrained discourse datasets, accessible right now, connected to clinical neuropsychological appraisals for dementia is the image depiction task assembled by the Alzheimer and Related Dementias Study at the University of Pittsburgh School of Medicine, regularly alluded to as "the Pitt dataset" [22], circulated through DementiaBank1. This dataset comprises of discourses from members who were recorded while playing out the Boston Cookie Theft picture portrayal task, from the Boston indicative aphasia assessment [23, 24, 25]. An assortment of computational strategies have been utilized on this corpus for discovery of Alzheimer's Disease and gentle intellectual hindrance across various investigations (more subtleties in segment 2). The majority of these works zeroed in on semantic provisions [26, 27, 28, 29], exploiting the manual records accessible with the discourse information. While paralinguistic highlights have so far got less consideration, there are valid justifications for examining a paralinguistics way to deal with AD. A portion of these reasons are methodological (for example staying away from the requirement for records) and some are identified with the idea of the sickness (for example the way that acoustic-prosodic investigation might prompt identification ahead of schedule and inconspicuous semantic decrease, yet additionally engine nuances with respect to discourse creation). Fraser et al. [29] were among quick to complete extra 1http://dementia. talkbank.org/acoustic-prosodic examinations on the Pitt accounts, extricating

42 mel-recurrence cepstral coefficient (MFCC) highlights [30], trailed by others [31, 32] which utilized a comparable methodology. Another new concentrate effectively utilized these accounts to remove low-level acoustic elements (vocalization occasions, discourse rate and number of expressions over a talk occasions) and utilized them to prepare a classifier for Alzheimer's patients and old controls [33]. This load of studies utilize the Pitt corpus, and most of their work depends on manual discourse records. Just [33] and the far reaching model by [29] present acoustic provisions, trailed and crafted by [31] and [32]. Nonetheless, these past examinations didn't change the dataset for expected confounders in an age and sex awkward nature or the impacts of variable sound quality, and utilized restricted specially appointed paralinguistic highlight sets. The work introduced in this paper resolves these issues by assessing an extensive arrangement of acoustic provisions which are arising in the field of computational paralinguistics, on a sexual orientation and age-adjusted subset of the Pitt corpus, which has been pre-handled to guarantee reliable sound quality.

The remaining article is organized as follows: part 2. Literature Review work 3. Section The proposed bidirectional LSTM network algorithm is presented in Section 3 follow by the Problem Statement. Section 4. Presents and discusses the results of the model simulation. Finally, in Section 5, the conclusions and suggestions are given for future work.

2. Literature Review

Lauraitis, A., Maskeliunas et al. [34] The learn contained 339 voice models assembled from 15 individuals: seven individuals with beginning stage CNSD (three Huntington's, one Parkinson's, one cerebral paralysis, one post-stroke, one early dementia), and 8 other tough individuals. The (Neural Impairment Test Suite-NITS) portable application's voice recorder is utilized to accumulate discourse information. Pitch shapes, Mel-recurrence cepstral coefficients (MFCC), Gamma tone cepstral coefficients (GTCC), Gabor (insightful Morlet) wavelet, and hear-able spectrograms are utilized to remove attributes. The accuracy of a sound versus debilitated order issue is 94.50% (Bi-LSTM) and 96.3% (WST-SVM). The innovation made could be utilized in computerized CNSD patient wellbeing state checking and clinical decision help plans, and likewise as a part of an (IoMT-Internet of Medical Things).

Rochford, I., Rapcan, V. et al. [35] The effect of applying stop and expression time sharing information in recognizing among the intellectually solid and weakened more seasoned people were concentrated in this review. Transient attributes with static 250 ms edge, brief qualities with dynamic limit, and respite and expression time assignment boundaries 3 arrangements of attributes were recovered from 187 catching discourses. Utilizing (LDA-

Linear Discriminant Analysis) classifications, the capacity of each one of these gatherings to recognize among intellectually solid and intellectually weakened members was attempted. Whenever differentiated to a static transient element, that noticed failure of a classifier utilizing a respite and the expression length circulation boundary upgraded by 0.22% (to 64.20% affectability), 6.33% (73.12% particularity), and 3.27% (68.66% complete accuracy).

Lopez-de-Ipina et al. [36] The goal of this article is to investigate the chance of utilizing keen algos to results obtained from non-intrusive scientific procedures on dubious patients to improve both early acknowledgment of Alzheimer's sickness or seriousness of an infection. This article assesses (ERAA-Emotional Response Automatic Analysis), which is subject to both conventional and novel discourse highlights: Higuchi (FD-Fractal Dimension) and Emotional Temperature (ET). The methodology has an unmistakable advantage of being, in adding to non-obtrusive, minimal expense, and liberated from angle impacts. This is a pre-clinical studio for approving forthcoming symptomatic tests and biomarkers. For a portrayal of qualities equipped for early analysis of Alzheimer's sickness, ERAA delivered amazingly astounding and confident results.

Mirheidari, B., Blackburn et al. [37] They include solid old controls (HCs) and those with MCI to a rundown of demonstrative classifications in this examination. They're likewise investigating whether IVA could be utilized to direct more customary intellectual evaluations, as verbal familiarity appraisals. A 4-way classifier arranged on an enormous list of capabilities acquired 48% accuracy, which expanded to 62% when just the 22 most significant provisions were utilized (ROC-AUC: 82%).

Luz, S. et al. [38] On an informational index of natural language catching of Alzheimer's patients (n = 214) and matured oversees (n = 184), a proposed method shows that a Bayesian classify working on qualities killed through simple algos for sound demonstrations acknowledgment and language rate catching could acquire an accuracy of 68%.

Liu, Z., Guo, Z., Ling et al. [39] The strategy for distinguishing dementia is examined in this paper by assessing an intuitive language made through Mandarin speakers during a picture depiction task. To start, a Mandarin discourse dataset is made, that incorporates a discourse from solid people and likewise patients by MCI (Mild Cognitive Impairment) or dementia. Earlier, 3 gatherings of qualities are recovered from voice accounts, including time attributes, acoustic attributes, and etymological qualities, and differentiated by making strategic relapse classifiers for dementia acknowledgment. Melding all qualities creates the best productivity for recognizing dementia from solid controls, with an accuracy of 81.9% in a 10-overlay cross-affirmation. Preliminaries are utilized to inspect the significance of different traits, and they show that fluctuation in perplexities delivered from phonetic models is the most helpful.

Haider, F., De La Fuente, S. et al. [40] From a computational paralinguistic point of view, investigation into the future worth of basically acoustic attributes naturally obtained from instinctual language for Alzheimer's dementia was distinguishing proof. On a likened model of Dementia Bank's Pitt instinctual language dataset, with patients compared through sex and period, a presentation of different cutting edge paralinguistic trademark sets for Alzheimer's ID was assessed. The (eGeMAPS-broadened Geneva moderate acoustic boundary set), an emblazoned trademark set, the differentiation 2013 trademark set, and the most recent Multi-Resolution Cochlea grams (MRCG) qualities were among the capabilities assessed. Besides, they additionally give the most recent (ADR-dynamic information portrayal) highlight extraction technique for Alzheimer's dementia recognition. The discoveries exhibit that an order structure dependent on acoustic discourse qualities extricated by means of our ADR strategy could acquire accuracy levels comparable to models utilizing more significant level language highlights. As per discoveries, all trademark sets contribute information that isn't gathered by other trademark sets. They exhibit that while the eGeMAPS property set offers marginally worked on exact (71.34%) than other trait sets actually, "hard combination" of trademark sets supports accuracy to 78.70%.

3. Proposed Work

3.1 Problem Identification

Support vector machine is used in which due to hyper plan accuracy was not sufficient. So, to increase the accuracy, efficiency and prediction rate we used deep learning.

3.2 Proposed Methodology

Dementia dataset is taken where the sound record is considered for discourse acknowledgment examination on the premise that information is produced and it is predefined as given in the dementia information databank. That sound record is changed over to message dependent on discourse investigation.

Then, at that point tokenization is performed on the information patient dataset. In this the dataset is cleant by eliminating and supplanting invalid strings and eliminating highlights. At long last, perform word tokenizing. Make age and sex records separately and produce a full information outline. Finally, we make a half breed model including1D-Conv and Bi-LSTM, then, at that point passes the information for preparing into the model.

Neural organization loads are saved and with the programmed discourse acknowledgment instrument dependent on NLP by perceiving a voice, it very well may be anticipated that the individual has dementia or not.

1) Data Preprocessing

Burden the dataset and convert it into a pickle record. Then, at that point perform tokenization on the information patient dataset. In this clean the dataset by eliminating and supplanting invalid strings and eliminating highlights. At long last, perform word tokenizing. Make age and sex records exclusively and create a full information outline. In this, it has two names dementia and control. Dementia comprises of 4 classes names that are treat, familiarity, review, and sentence.

Presently Encode the tokenized information utilizing UTF-8 that comprises of a tokenized list and tokenized id that makes an information outline utilizing message, level, and id in the dementia list. This information outline saves in CSV design. Presently measure the Ana realistic record that comprises of id, section age, introductory date. Then, at that point make a patient word reference by thinking about id, age, sex, race, and training. Presently make the last information outline including all ascribes from Ana's realistic document and patient word reference. Presently examine the power of Sentiments by converging up the pickle record and the last information outline that is in CSV. Then, at that point we characterized the Post tag in preprocessing, then, at that point additionally considered jargon size of 30000, the grouping length of 73, and installing size of 300.

2) Preparing

Pass pickle record for preparing and figure post labels for all meeting documents and split the dataset train 90% and test 10% with 4, 10, and 95 irregular seeds. Presently perform word inserting utilizing Glove6V. These are utilizing an Ada-Gard enhancer with a learning cost of 0.0001 and at long last, set up the organization.

Bidirectional LSTM

Bidirectional long-present moment memory (bi-lstm) is the most common way of making any neural organization have the succession data in the two ways in reverse (future to past) or forward (past to future).

In bidirectional, our feedback streams in two ways, making a bi-lstm unique in relation to the standard LSTM. With the normal LSTM, we can make an input stream one way, either in reverse or forward. Notwithstanding, in bi-directional, we can make the info stream in the two ways to protect the future and the previous data.

In the chart, we can see the progression of data from in reverse and forward layers. BI-LSTM is normally utilized where the arrangement to successive

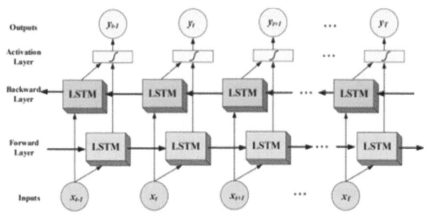

Fig. 3.2.1. Bi-LSTM model.

assignments is required. This sort of organization can be utilized in text order, discourse acknowledgment and determining models.

Bidirectional LSTMs are an augmentation of customary LSTM which could be utilized to build model proficiency on grouping classification issues.

Bidirectional LSTMs plan 2 LSTMs on an information succession rather than one in issues where unequaled strides of an information series are available. The first on an info succession, and a second on a turned around reproduction of it. This could offer the organization more setting and lead to faster and significantly more learning of an issue.

In this paper, the Summary of the Bi-LSTM model is clarified layer-wise.

3) Add Model layer

Bidirectional LSTM

We can stretch out a case to exhibit a Bidirectional LSTM since we realize how to make a LSTM for arrangement order issues.

This should be possible by wrapping a LSTM discharged surface in a Bidirectional surface, as seen beneath: This will build 2 duplicates of an emitted surface, one that will fit in an information series with no guarantees and one a switched duplicate of an info series.

Info layer: input sentence to this system.

Installing layer: Map each word into a lower measurement vector.

Attention layer

Make a weight vector, and increase it by a weight vector to consolidate word-level attributes from each time venture into a sentence-level trademark vector.

Dropout

Dropouts are a regularization procedure that is utilized to keep away from a system from overfitting. Dropouts are utilized to change a level of an organization's neurons at irregular intervals. The arriving & leaving connect to those neurons and are likewise turned off when neurons are turned off. This is finished to assist a system with learning. Dropouts ought not be used after convolution layers; all things considered, they should be used after the organization's thick layers. This is consistently an extraordinary thought to just mood killer neurons to half. On the off chance that we turn off the greater part of a structure, quite possibly the system might incline seriously & a conjecture will be mistaken. How about we see how to utilize dropouts and portray them while making a Bidirectional LSTM Model.

Dense

The term alludes to which neurons in an organization surface are completely connected (thick). Each neuron in a surface before it accumulates information from all neurons in a surface before it, making them amazingly interconnected. In different terms, a thick surface is a completely interconnected surface, demonstrating that all neurons in 1 layer are coupled to those in the following.

3.3 Proposed Algorithm

A. Distributed computing advantages and downsides design of a distributed computing can be classified into four layers:

The Physical layer, the foundation layer, the stage layer, and the application layer, as shown in Figure.

Step1: Collect the information Dementia dataset

Step2: Audio document is considered for discourse acknowledgment

Step3: After that is created and predefined as given in the dementia bank.

Step4: After that the sound document is changed over to message dependent on discourse examination utilizing NLP.

Step5: Then perform tokenization on the information patient dataset.

Step6: In this clean the dataset by eliminating and supplanting the invalid strings, accentuation, and eliminating highlights. At last, perform word tokenizing. Make age and sex documents independently and produce a full information outline.

Step7: Now Encode the tokenized information utilizing UTF-8 that comprises of a tokenized list and tokenized id that makes an information outline utilizing text, level, and id in the dementia list.

Step8: Create a half breed model including 1D-Conv and Bi-LSTM, then, at that point pass the information for preparing.

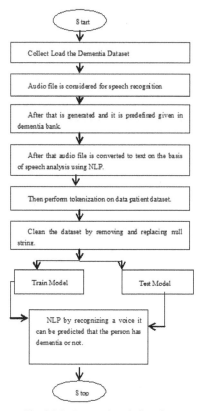

Fig. 3.3.1. Proposed work flowchart.

Step9: Finally deliberate the model exhibition on test information estimated from test information in a part of exactness, accuracy, review, and F-score. After finishing the preparation execute the model.

Step10: Neural organization loads are saved and with the programmed discourse acknowledgment apparatus dependent on NLP by perceiving a voice it very well may anticipate that the individual has dementia or not.

4. Results and Discussion

This work has been implemented using Python programming language and the platform used is Jupyter notebook (version 6.3.1). here, we have used the Dementia dataset. Experiment. The description of such a dataset and achieved results of the proposed model are given below.

4.1 Dataset Description

Dataset Dementia Bank (Boller and Becker 2005) is the biggest publicly available dataset of transcripts.

Voice recordings of AD interviews (& control) patients. 1. Patients were required to complete several activities, like the "Boston Cookie Theft" explanation activity, patients were shown an image & required them to explain what they saw (See Figure 4). Other activities included the 'Recollect Test,' in which patients were asked to recall details from a previously stated story. Automatic morph syntactic analysis, like the standard part-of-speech labeling, explanation of tense, & repetition indicators, is included with each transcript in the Dementia Bank. 2. Note that these are not AD-specific characteristics, but rather generic, automatically extracted language qualities. To use as datasets, we separated every transcript into individual utterances. We also deleted any utterances that were not accompanied by POS tags. This balancing lowered the amount of data but assured that models with tagged & untagged settings were matched fairly.

The Alzheimer & associated Dementias Research at the University of Pittsburgh School of Medicine obtained these transcripts & audio files as part of a wider protocol. The University of Pittsburgh was awarded NIH funds AG005133 & AG003705 to help them acquire Dementia Bank information. Individuals with probable & suspected Alzheimer's Illness, as well as elderly controls, took part in a study. Data was collected every year throughout a period.

https://dementia.talkbank.org/access/English/Pitt.html

Fig. 4.1. Boston cookie stealing task description. All activities in the picture were to be explained by participants.

4.2 Performance Matrix

1. Accuracy

The number of correct forecasts your model made for an entire test dataset is referred to as accuracy. The following formula is used to calculate it:

$$Accuracy = \frac{TP + TN}{TP + TN + FP + FN} \tag{1}$$

2. Recall

The right positive price, also called recall, is an evaluation of how many right positives are forecasted out of all positives in a dataset. Sensitivity is another name for it. The following method is used to measure a metric:

$$Recall = \frac{TP}{TP + FN} \tag{2}$$

3. Precision

Accuracy is a criterion for evaluating how precise a positive forecast is. In other terms, how sure could you be that a positive result is indeed positive if it is forecasted as such? It is evaluated using the formula below:

$$Precision = \frac{TP}{TP + FP} \tag{3}$$

4. F-score

The F1-score is by far a most popular F-score. It is their harmonic mean, which is a mixture of precision & recall. F1-score could be measured using the formula below:

$$F1 = 2 \cdot \frac{Precision.recall}{Precision + recall} \tag{4}$$

4.3 Outcome of the Model

The comparison of several classification methods is represented in Table 4.3.1. It represents overall performance comparison output in contrast to several existing methods and Bi-LSTM Method like Accuracy, Precision, Recalls, and F1-Score.

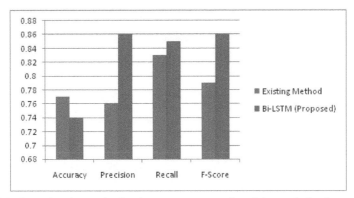

Fig. 4.3.1. Comparison bar graph of performance parameters for existing method and proposed and Bi-LSTM.

Table 4.3.1. Comparison of performs parameter between existing method and Bi-LSTM (Proposed) method.

Parameter	Existing Method	Bi-LSTM (Proposed)
Accuracy	0.77	0.74
Precision	0.76	0.86
Recall	0.83	0.85
F-Score	0.79	0.86

5. Conclusion

Dementia ailment impacts discourse is a kind of syntactic, semantic, information, and hearable issue, as indicated by results from past investigations. Without utilizing master characterized phonetic components, we utilized an exchange learning methodology to improve independent AD estimating utilizing a nearly minimal centered talking dataset. On a Cookie-Theft picture clarification check of a Pitt corpus, we tried a recently built pre-arranged transformer reliant upon talking structure which we improved with upgrade techniques. The exactness (74%), accuracy (86%), review (85%), and F1 scores of (86%) were gotten utilizing sentence level Bi-LSTM, which improved cutting edge results. To increase the performance parameter some more models can be added along with Bi-LSTM. In a few dialects, pre-prepared language models are open. As a result, a strategy introduced in this review could be tried in dialects other than English. Moreover, with multilingual forms of these plans, data of AD determination in 1 language could be communicated to another dialect if a sufficiently large dataset isn't available. In future, we might accumulate a bigger dataset that might help in the formation of a more summed up implanting. Further, we can likewise expand the dataset for individuals communicating in various dialects.

References

[1] American Psychiatric Association. 2000. Delirium, dementia, and amnestic and other cognitive disorders. In Diagnostic and Statistical Manual of Mental Disorders, Text Revision (DSM-IV-TR), ch. 2.

[2] Burns, A. and Iliffe, S. 2009. Dementia. BMJ, vol. 338.

[3] World Health Organization. 2013. Mental health action plan 2013–2020. WHO Library Cataloguing-in-Publication DataLibrary Cataloguing-in-Publication Data, pp. 1–44.

[4] World Health Organization. 2015. First WHO ministerial conference on global action against dementia: meeting report. WHO Library Cataloguing-in-Publication DataLibrary Cataloguing-in-Publication Data, pp. 1–76.

[5] Maccioni, R. B., Far´ıas, G., Morales, I. and Navarrete, L. 2010. The revitalized tau hypothesis on Alzheimer's disease. Archives of Medical Research 41(3): 226–231.

[6] Hardy, J. and Selkoe, D. J. 2002. The amyloid hypothesis of Alzheimer's disease: progress and problems on the road to therapeutics. Science 297(5580): 353–356.

[7] Braak, H. and Braak, E. 1996. Evolution of the neuropathology of Alzheimer's disease. Acta Neurologica Scandinavica 94: 3–12.

[8] Norton, S., Matthews, F. E., Barnes, D. E., Yaffe, K., Brayne, C. et al. 2014. Potential for primary prevention of Alzheimer's disease: an analysis of population-based data. The Lancet Neurology 13(8): 788–794.

[9] Mendez, M. F., Cummings, J. L. and Cummings, J. L. 2003. Dementia: A Clinical Approach (3rd edition). Butterworth-Heinemann.

[10] Ross, G. W., Cummings, J. L. and Benson, D. F. 1990. Speech and language alterations in dementia syndromes: Characteristics and treatment. Aphasiology 4(4): 339–352.

[11] Kirshner, H. S. 2012. Primary progressive aphasia and alzheimer's disease: brief history, recent evidence. Current Neurology and Neuroscience Reports 12(6): 709–714.

[12] Bondi, M. W., Salmon, D. P. and Kaszniak, A. W. 1996. The neuropsychology of dementia. In Neuropsychological Assessment of Neuropsychiatric Disorders, 2nd ed., pp. 164–199.

[13] Dawadi, P. N., Cook, D. J. and Schmitter-Edgecombe, M. 2013. Automated cognitive health assessment using smart home monitoring of complex tasks. IEEE Trans. Syst., Man, Cybern., Syst. 43(6): 1302–1313.

[14] Braaten, A. J., Parsons, T. D., Mccue, R., Sellers, A. and Burns, W. J. et al. 2006. Neurocognitive differential diagnosis of dementing diseases: alzheimer's dementia, vascular dementia, frontotemporal dementia, and major depressive disorder. International Journal of Neuroscience 116(11): 1271–1293.

[15] Folstein, M. F., Folstein, S. E. and McHugh, P. R. 1975. Mini-mental state. A practical method for grading the cognitive state of patients for the clinician. J. of Psychiatric Res. 12(3): 189–98.

[16] Robert, P., Schuck, S., Dubois, B., Lepine, J., Gallarda, T. et al. June 2003. Validation of the Short Cognitive Battery (B2C). Value in screening for Alzheimer's disease and depressive disorders in psychiatric practice. Encephale 29(3 Pt 1): 266–72.

[17] Dubois, B., Slachevsky, A., Litvan, I. and Pillon, B. 2000. The FAB: a frontal assessment battery at bedside. Neurology 55: 1621–6.

[18] Mathuranath, P. S., George, A., Cherian, P. J., Mathew, R., Sarma, P. S. et al. 2005. Instrumental activities of daily living scale for dementia screening in elderly people. Intl. Psychogeriatrics 17(3): 461–74.

[19] Forbes-McKay, K. E. and Venneri, A. 2005. Detecting subtle spontaneous language decline in early Alzheimer's disease with a picture description task. Neurological Sciences 26(4): 243–254.

[20] Konig, A., Satt, A., Sorin, A., Hoory, R., Toledo-Ronen, O. et al. 2019. Automatic speech analysis for the assessment of patients with predementia and 1932–4553 (c) 2019 IEEE.

[21] Konig, A., Linz, N., Tr"oger, J., Wolters, M., Alexandersson, J. et al. 2018. Fully automatic speech-based analysis of the semantic verbal fluency task. Dementia and Geriatric Cognitive Disorders 45(3-4): 198–209.

[22] Becker, J. T., Boiler, F., Lopez, O. L., Saxton, J., McGonigle, K. L. et al. 1994. The natural history of alzheimer's disease. Archives of Neurology 51(6): 585.

[23] Goodglass, H. and Kaplan, E. 1983. The assessment of aphasia and related disorders. Philadelphia.

[24] Goodglass, H., Kaplan, E. and Barresi, B. 2001. The Assessment of Aphasia and Related Disorders. Lippincott Williams & Wilkins.

[25] Goodglass, H. 2000. Boston Diagnostic Aphasia Examination: Short Form Record Booklet. Lippincott Williams & Wilkins.

[26] Orimaye, S. O., Wong, J. S., Golden, K. J., Wong, C. P., Soyiri, I. N. et al. 2017. Predicting probable Alzheimers disease using linguistic deficits and biomarkers. BMC Bioinformatics 18(1): 34.

[27] Orimaye, S. O., Wong, J. S.-M. and Wong, C. P. 2018. Deep language space neural network for classifying mild cognitive impairment and Alzheimer-type dementia. PloS One 13(11): e0205636.

[28] Fraser, K. C., Fors, K. L. and Kokkinakis, D. 2019. Multilingual word embeddings for the assessment of narrative speech in mild cognitive impairment. Computer Speech & Language 53: 121–139.

[29] Fraser, K. C., Meltzer, J. A. and Rudzicz, F. 2016. Linguistic features identify Alzheimer's disease in narrative speech. Journal of Alzheimer's Disease 49(2): 407–422.

[30] Chen, J., Wang, Y. and Wang, D. 2014. A feature study for classification based speech separation at low signal-to-noise ratios. IEEE/ACM Trans. Audio, Speech, Language Process 22(12): 1993–2002.

[31] Yancheva, M. and Rudzicz, F. 2016. Vector-space topic models for detecting Alzheimers disease. In Procs. of ACL, pp. 2337–2346.

[32] Hernández-Domínguez, L., Ratte, S., Sierra-Martínez, G. and RocheBergua, A. 2018. Computer-based evaluation of Alzheimers disease and mild cognitive impairment patients during a picture description task. Alzheimer's & Dementia: Diagn., Assessm. & Dis. Monit. 10: 260–268. doi: 10.1016/j.dadm.2018.02.004.

[33] Luz, S. 2017. Longitudinal monitoring and detection of Alzheimer's type dementia from spontaneous speech data. In Procs. of the Intl. Symp on Comp. Based Medical Systems (CBMS). IEEE, pp. 45–46.

[34] Lauraitis, A., Maskeliunas, R., Damasevicius, R. and Krilavicius, T. 2020. Detection of speech impairments using cepstrum, auditory spectrogram and wavelet time scattering domain features. IEEE Access 8: 96162–96172. doi:10.1109/access.2020.2995737.

[35] Rochford, I., Rapcan, V., D'Arcy, S. and Reilly, R. B. 2012. Dynamic minimum pause threshold estimation for speech analysis in studies of cognitive function in ageing. 2012 Annual International Conference of the IEEE Engineering in Medicine and Biology Society. doi:10.1109/embc.2012.6346770.

[36] Lopez-de-Ipina, K., Alonso, J. B., Travieso, C. M., Egiraun, H., Ecay, M. et al. 2013. Automatic analysis of emotional response based on non-linear speech modeling oriented to Alzheimer disease diagnosis. 2013 IEEE 17th International Conference on Intelligent Engineering Systems (INES). doi:10.1109/ines.2013.6632783.

[37] Mirheidari, B., Blackburn, D., OrMalley, R., Walker, T., Venneri, A. et al. 2019. Computational cognitive assessment: investigating the use of an intelligent virtual agent for the detection of early signs of dementia. ICASSP 2019 - 2019 IEEE International

Conference on Acoustics, Speech and Signal Processing (ICASSP). doi:10.1109/icassp.2019.8682423.

[38] Mirheidari, B., Blackburn, D., OrMalley, R., Walker, T., Venneri, A. et al. 2019. Computational cognitive assessment: investigating the use of an intelligent virtual agent for the detection of early signs of dementia. ICASSP 2019 - 2019 IEEE International Conference on Acoustics, Speech and Signal Processing (ICASSP). doi:10.1109/icassp.2019.8682423.

[39] Haider, F., de la Fuente, S. and Luz, S. 2020. An assessment of paralinguistic acoustic features for detection of alzheimer's dementia in spontaneous speech. In IEEE Journal of Selected Topics in Signal Processing 14(2): 272–281, Feb. 2020, doi: 10.1109/JSTSP.2019.2955022.

[40] Haider, F., De La Fuente, S. and Luz, S. 2019. An assessment of paralinguistic acoustic features for detection of alzheimer's dementia in spontaneous speech. IEEE Journal of Selected Topics in Signal Processing, pp. 1–1. doi:10.1109/jstsp.2019.2955022.

Index